John Hartig

The Universe Explained
Well, sort of…

Copyright © 2024
Book cover design and writing
John Hartig

Published on Amazon and Kindle
with thanks for the chance
to publish affordably
and for the helpful phone chats

Part One

Part One	
Topics Discussed	
Where to Begin	p. 2
Remembrance Day	p. 5
Consciousness in Everything	p. 9
The Evolutionary Brain	p.12
The Creationist Brain	p. 16
Stephen Hawking	p. 19
Perspective	p. 24
God or Nothing	p. 28
How Big is Big	p. 33
Finite versus Infinite	p. 35

Where to Begin?

When I was a kid, I used the Encyclopedia Britannica to look up information. The poor man's version was, of course, Funk and Wagnalls. Nowadays, it's Google and YouTube.

I found YouTube especially useful in constructing this book. I summarized the contents of videos on science and outer space, hoping to simplify them for the average person. The intention of this book is to widen an interest in science ever since the electron microscope and the James Webb Telescope have given us more knowledge about what things are made of.

We have amazing photos of galaxies which were formed almost 13.7 years back in time. How were stars made? Are there things smaller than electrons? Why haven't we discovered aliens? Did God create the Big Bang? What is the meaning of life?

I hope you find my summaries of YouTube videos instructive. You will be amazed at the scientific questions astronomers and physicists are now asking.

Einstein looked for mathematics, a formula, that would explain a Unified Theory of Everything. He never succeeded. I've attempted in words to explain the whole of Existence in my book, or at least, that you get an inkling of Everything. YouTube makes amazing videos and documentaries available to the public to educate them on the latest discoveries in star study and in physics. We have boundless theories about what makes our universe tick and the star stuff we are made of.

My book will help you in these areas of thought because decade by decade we are living in an amazing time of new discoveries. There is a lot to unfold from the first development of civilization to our present level when we can reach the stars.

Where did our gift of reason come from? It's even speculated that extraterrestrials, the Anunnaki, from the planet Nibiru, genetically engineered us and taught us building and writing. Others say that God created everything, and a savior had to be sent to mankind to bring us back to God. Yet again, others say that we are the product of biogenesis, of natural processes in nature which took billions of years to evolve us.

It is no easy task to wrap up 13.8 billion years in plain language so that evolution, consciousness and God can be understood. Perhaps, there is something to the concept of the Kardashev Scale. Come with me along this journey where we speculate about the size of galaxies, clusters of galaxies and the possibility of a conscious universe.

Max Bennett, an artificial Intelligence expert, wrote <u>A History of Intelligence: Evolution, AI, and the Five Breakthroughs That Made Our Brains</u>, in 2023. I found this book as interesting as Stephen Hawking's, <u>A Brief History of Time</u>, published in 1988.

*****

Ironically, as I begin my book, it is November 11th, 2024, Remembrance Day! Did the dead ever wonder what they were fighting for, what life was all about? Did we have a purpose in the universe or did the universe have a purpose for them?

Since we are doing our best to destroy each other in regional wars, one wonders if we had any brains at all. So many of our ancestors have died in the previous two world wars to "fight" for peace, and to die for freedom and democracy.

Aliens might look at us and decide that intelligence might be found on another planet, not here!

Remembrance Day

Yes! November 11, 2024, the 11th day of the 11th month at the 11th hour is a good time to start my new book to explain everything, including the meaning of life.

Donald Trump is president again and wars still rage in Gaza and Ukraine. What would Carl Sagan, the great astrophysicist, think about our Earth as he looked back on this pale blue dot seen from 3.7 billion miles within deep space.

If we are the only civilization within the Milky Way Galaxy, perhaps within the whole universe, then we have a huge responsibility because without us, there would be no meaning to anything. Carl Sagan appreciated our preciousness in the photograph of a pale blue dot taken by Voyager 1 on February 14th, 1990.

> To my mind, there is perhaps no better demonstration of the folly of human conceits than this distant image of our tiny world. To me, it underscores our responsibility to deal more kindly and compassionately with one another and to preserve and cherish that pale blue dot, the only home we've ever known.

What a waste, what a tragedy war is! I'm reminded of the huge waste in life, of talent and splendid potential, in all the wars before 1914 and thereafter. George Butterworth, a wonderful composer, shot in the head on

August 5th, 1916, best known for the orchestral idyll, *The Banks of Green Willow*. Thank goodness that one of my other favourite composers, Ralph Vaughan Williams, escaped such a tragedy when doing his duty as an ambulance driver. He composed one of my most enjoyable tunes, *The Lark Ascending*, in 1914 during the early days of WW 1. I think of the birds who fly high in circles above the deadly skirmishes which humans wage with their guns and cannons so far below, making destructive noises that drown out the sweet twitter of fragile birds.

I yield two minutes of silence every year to the memory of Lieutenant Colonel John McCrae, MD, and all the fallen soldiers on both sides. John McCrae has always meant something special to me since I moved to Canada from a war-torn Austria in 1954. He wrote that famous poem, *In Flanders' Fields*.

> In Flanders fields the poppies blow
> Between the crosses, row on row,
> That mark our place; and in the sky
> The larks, still bravely singing, fly
> Scarce heard amid the guns below.
> https://en.wikipedia.org/wiki/John_McCrae

John McCrae died near the end of WW 1 of pneumonia at the age of 45. He was a physician, poet, author and artist, another example of the loss of a valuable mind! I mentioned him, Ralph Vaughan Williams and George Butterworth in my memoir of a refugee, You Love Our Milk and Honey, John Hartig, 2020 publ. through Amazon.

Imagine, if you will, looking back at the pale blue dot of Earth from 3.7 billion miles out in deep space, how insignificant we are within that huge vastness and what a shame it would be if we destroyed ourselves, the source of meaning to everything in the universe, unless the universe knows itself! Maybe that is so; scientists have talked about the universe's evolution into a living, thinking, knowing entity.

Anyway, astronomers have mapped out a timeline which traces the history of the universe back in time to the Big Bang some 13.82 billion years ago. The cosmos was made up of whizzing quarks, neutrons and electrons in an immense hot and dense state in that first one-tenth-millionth-of a trillionth-of a trillionth of a second. After this fraction of time, the four forces went their separate ways, and the first subatomic particles were able to hold themselves together. At T+20 minutes, the universe made atoms, isotopes of hydrogen, helium and lithium and a smattering of beryllium. After 380,000 years passed, the universe entered the phase of recombination where free electrons combined to form photons. During its ionized stage, the universe was dark; when recombination happened, light happened. This is the light of which we have the imprint today as the microwave background. The temperature of the microwave background is a consistent 2.725 Kelvin. According to astrophysicist, Alan Guth, "inflation" accounted for the super-expansion of the universe which stretched some lumpy irregularities throughout which turned into our stars, planets and galaxies.

Some intriguing questions are being investigated:
- What is dark energy?
- What is dark matter?
- Where did the universe come from in the first place?
- Are there more universes out there?

One would expect a book on the universe to start at the beginning of time, at the beginning of existence, some

13.8 billion years ago when "the primordial atom" exploded in a Big Bang to create everything. But you already know all that since Stephen Hawking planted the seeds of that knowledge in his explanation of <u>A Brief History of Time.</u> By that he meant 13.8 billion years ago!

A thought passed my mind, that the primordial atom wasn't an atom at all. It might have been a super primordial black hole, or white hole for that matter, which exploded into our existence.

Questions have been asked about what happened before the Big Bang, but then, since time was created at the beginning of existence, it does not make sense to ask such a question because that is like asking what is north of the north pole? Actually, the astrophysicist Roger Penrose has an answer for that in his cyclical model of a repeating universe, so there might have been a time before time, which he calls eons.

However, starting my book on Remembrance Day is indeed fitting [before we get sidetracked] because the intention of this book is to make us think about what on earth we are doing to our planet! We are defiling it with our garbage, our plastics, our smoke and climate change and above all, bombing each other in regional wars. We need to stop our stupidity!

So, where did our reasoning and our consciousness come from? Because an alien might have doubts about whether we have any!

Consciousness in Everything!

It's even been posited by astrophysicists that we live in a "simulation", something that a giant computer can create like a world within Star-Trek's holodeck. Astrophysicists touch on metaphysics here, the realm of guessing and of imagination, like the old TV series, *The Twilight Zone*.

Why Consciousness is Immortal | The Philosophical Proof of Life After Death
https://www.youtube.com/watch?v=DNGT0uYPHAo&t=47s

Questions here are permissible like, is there life after death; is life real; are we immortal? If the whole universe is part of evolution, then it has an instinct for self-preservation and survival. This opens the door to the paranormal and to other dimensions where beings invisible to us live in their own worlds. This is an area beyond my comprehension. The topic escapes me and I fear I do not have the intelligence to synopsize Thomas Clark's arguments when I do not understand him to begin with. I don't feel comfortable in this area of metaphysics.

Anyway, Philosopher Thomas Clark wrote an essay on "Death, Nothingness and Subjectivity." Clark claims that we are "that part of reality that becomes aware of its existence. For better or worse, we are truly immortal. We can experience death but only as an abstraction...from the outside. But it is an illusion. There is no life after death, because there is no death." Tell that to my dearly departed friends!

This I confess, escapes my understanding, because there is no proof, though it might make great discussion in a university coffee shop. The smartest geniuses who ever lived, like William James Sidis, with an IQ of 250 might have trouble digesting the ideas broached by such philosophers as Thomas Clark. I do not like to get into philosophical gobbledygook anyway in my book, so I will not waste my time in this area of "metaphysical speculation", trying to explain that which I cannot explain.

The Living Universe-Documentary about Consciousness and Reality
https://www.youtube.com/watch?v=HD4WthE414k

Since the time of Plato, philosophers have entertained the idea that the universe is a great mind, while others think it is a great machine. Panpsychism claims that the universe is able to be aware. It is argued that life and consciousness are somehow necessary, value-giving features of reality, that ultimately contribute in some mysterious way to the metaphysical integrity of existence itself. Any value is unintelligible in a universe without mind and several contemporary philosophers have argued that it is not unreasonable to consider that some primordial notion of value is required to explain the universe's existence. A universe in the far future may evolve into a self-aware entity. Again, this region of speculation is just that, metaphysical fascination, speculation without any proof.

Scientists Say the Universe is in Someone's Brain
https://www.youtube.com/watch?v=8v4ONqxzzjk
Could we be like cells in a giant creature but on a planetary scale that hasn't realized its existence yet? How could we even find out? That the universe on a large scale is similar to the construction of our brain raises the question of whether the universe has similar thinking capacities. In 500 B.C. Annexures posited that the universe was a living thing. A smart force, naous, guides the universe to become more organized and purposeful. That's evolution, isn't it?

In August 2022, theoretical physicist, Sabine Hossenfelder, wrote an article claiming that maybe the universe thinks! She points out the similarities between the structure of the universe and the human nervous system. The galaxies are bound together by gravity into clusters which then form into larger clusters. These clusters are connected by galactic filaments, long threads of galaxies, which form dendritic networks like the neurons in the human brain, with vast voids punctuating the spaces between them creating an awe-inspiring celestial web. In the brain, neurons also cluster together forming larger groups. There is a parallel between the cosmic web and human brain's web. Therefore, the question, is the universe conscious? Panpsychism suggests that the tiniest bits of matter have a conscious experience. Some scientists have said that consciousness is an inherent property of matter in varying degrees. Maybe that is why the microscopic world of quantum mechanics is so elusive.

The Evolutionary Brain

    I remember my grade 8 IQ test. I don't know how well I did except it upset me. I had to do it twice, by the way. I was a foreign kid from Austria after WW II ended, maybe tackling an IQ test in English was the glitch. I'm sure IQ tests in Canada were biased toward the English-speaking culture. Anyway, I ended up in a streamed class the next year anyway going into grade 9A in an all-boys Catholic High-School. The A class was achievers.

    There is a difference between achievers and the gifted. I never considered myself as gifted. I had to wander back and forth in my room to memorize things for exams by repeating them countless times. Jim, a classmate of mine, absorbed things like a sponge. He could repeat a whole page in a book without blinking. The brain, whether ordinary or extraordinary, is an extraordinary organ, nevertheless. Where did our capacity for imagination and reasoning come from? Where did our memory and consciousness come from?

*****

Bl 181 Max Bennett: A Brief History of Intelligence
https://www.youtube.com/watch?v=Gt386yZZx5I
Dec. 25, 2023

    Max Bennett sees the brain as a three-fold organ just like neuroscientist Paul MacLean does. First is the reptilian part in the brain stem which is the seat of our survival instincts. A sequential addition which mammals have is the limbic component. Humans have a neocortex which gives imagination and reasoning. This theory is replaced now by what neuroscientists call the adaptive brain. However, there is no map so to speak with which you can reverse engineering the brain and apply it to artificial intelligence. Bennett looks at modifications which the brain has undergone and how this can be applied to Ai. Bennett says that mammals have evolved different capacities, which evolved in a specific order. "We understand what capacity algorithms work and do not work." He looks for "first approximations" where mammals are able to imagine futures with episodic like memory and counterfactual learning of what works and doesn't work. He sees this first approximation evolving out of the ability to simulate. That way mammals can imagine the state of the world that is not the current one. Bennett says that this capacity to imagine futures, rendering pasts and considering alternative past choices is an important step in the evolution of intelligence. Bennett sees this first approximation as a 5-step breakthrough in brain Evolution. Part of this evolution involves:

- Steering or navigating the world without understanding the world...like a "nematode" which cannot see, except they have photosensitive cells...Bennett says, "What is beautiful about life in the universe is how it's founded on randomness and yet common things are repeatedly recapitulated which is a beautiful fact about life..."
- Primates have the ability to imagine the self...they can also imagine other people's perspective, so they are self-aware...this maps to the idea of mentalizing or imitation learning...
- Anticipating future needs...humans have the ability to imagine future need states...

Bennett was asked if we could improve the human brain, knowing what we know about its make-up today. Bennett says that mentalizing would be an important transfer to give to Artificial Intelligence, i.e., AI. The ability to ensure that AI understands what we mean when we say something...as in the Paperclip Problem where a robot is told to maximize the production of paperclips and goes ahead to use the whole world to do so, so that the entire world is used for nothing but paperclips...Mentalizing or imagining what the boss means by what he says is an important inference...Mentalizing the difference between sensible action and ridiculous extremes...words for an AI must involve sensible inference and not lead to misinterpretation. Inference is a higher function of the brain.

Bennett does not define intelligence in his book, noting that there are different intelligences and capacities to learn. Bennett posits an interesting question about intelligence: if defined as the ability to solve problems to achieve goals. He sees it silly if a goal is the objective, especially in terms of the universe. What goal does the universe have, except entropy? Another definition claims that intelligence involves learning, which isn't always the case. The definition of intelligence has a lot of amorphous boundaries. Look at the "idiot" who plays the banjo in the movie "Deliverance".

Language is the 5th breakthrough which the brain has mastered. Bennett says it enables interbrain transfer of ideas which is a powerful evolutionary tool.

What does the future hold for brain evolution? Obviously to pass our intelligence on to AI! Bennett sees three outcomes:
- That humans subdue AI through regulation so that AI remains as a computerized tool.

- Another possibility is that humans are supplanted by Ais. The future might be one where parents do not choose to have children. Bennett sees this choice as likely to have a Human/AI system.
- The third alternative is merging between humans and Ais. Cyborg hybrids. Bennett sees this as too costly. He doesn't see a future where you push a button, and you have French uploaded to your brain as a new skill.

Max Bennett sees his book as piecing together the ideas of other experts. He does his best to give them credit where credit is due. He asserts that he did not have an ulterior motive for a career move in academia. He says that his book was instructive and worth the effort where he had to write with clarity so that a wider audience could appreciate his effort. In a way, my goals in my own book are similar.

The Creationist Brain

Creationists have often been accused of taking the easy way out by saying that intelligence can only come from an intelligent source. It takes a brain to create a brain.

The Brain Didn't Evolve – A Psychiatrist Explains Why
https://www.youtube.com/watch?v=wkodiXV3KHQ

Psychiatrist Dianne Grocott contends that the brain did not evolve, contrary to Max Bennett's premise, but that it had to be created by intelligent design. The brain has 86 billion neurons so that such complexity could only come from a Creator. She sees the atheistic view relying on 5 points or pillars, which she sees as flimsy arguments:

- An atheist would argue that everything came out of nothing. Dr. Grocott argues that you needed a hand "to light the match" of the first light.
- Next is chemical evolution. Non-life has become life which does not make sense to her because life needs order, information and design which has to come from somewhere.
- Biological evolution where microbes become monkeys and monkeys become men. DNA is too complex to accept that jump in changes. To put in one extra gene, it would take the whole time of the universe to do that which is 13.8 billion years. "It's not possible; monkeys did not turn into humans."

- The 4th point is a misunderstanding about what natural selection is. Breeding different selections of dogs is natural selection but it is not evolution. A dog will remain a dog and not turn into anything else. To get different species, however, there has to be a Creator.
- The 5th point is that we can choose the glasses through which we can look at the data. "I think that is evidence for design." She said she became an atheist for a while. "I had a choice; I put on those glasses." At about the age of 30, she met God. "I was surprised. You'd been there all the time. And you're actually nicer than I thought…My eyes were open. It's like you put on clean glasses."

Dr. Dianne Grocott says that the brain is so logically organized into regions that it could only have been designed by God. The frontal lobe is your CEO; your temporal lobes are at your temple, responsible for your memory; your parietal lobes are at the top, giving you spatial awareness and control of your movements; finally, the back lobes are what she calls your visual arts studio for organizing what you can see; the limbic system is in the middle for regulating your emotions and survival. Dr. Grocott considered the brain stem as boring, although it's the system that keeps you alive through instinct.

All the regions communicate with each other through the synapses which she compares to your social media. "They chat among themselves and use neurotransmitters, little chemicals, as well as larger cells that are like cables to create a sophisticated communication system." She is amazed at what researchers have discovered in the workings of the brain, especially to help people whose brains are diseased.

The brain has miraculous neuroplasticity so that half the brain can be removed when a person is a child, and the brain seems to compensate in most areas. The same compensation does not take place, however, in an older person who had a stroke. Yet, Dr. Grocott sees every level of the brain as so intricately arranged that the whole construction has to be divinely designed. She lists these abilities: language, symbolism, imagining possible futures, abstract thought, altruism, capacity for music, and spiritual awareness. All of this makes us human, not coming from evolution but God's Design, she says. She claims that God loves us, like we love our kids. "But if everything evolved, you don't need families!" She adds, "The Designer allows us to say, I don't want to know you, but there are consequences."

Personally, I feel uncomfortable with Dr. Grocott's enthusiasm in her argument that the brain was created by intelligent design. She seems to be enthralled by the wonder of creation, so her wonder reminds me of a religious zealot.

My opinion is that the brain was the product of evolution and that God, since I am a Chistian, intervened at points to supervise the whole evolution process. I also do not believe that the world is 6,000 years old as fundamentalists claim. I subscribe to the idea that amoebas and dinosaurs preceded us and that the universe is 13.7 billion years old, and the Earth itself is 4.543 billion years old. That evolution was the process which brought everything through natural phases, guided by the power of God.

A Brief History of Time: The Pioneering Work of Stephen Hawking | Naked Science | Spark

https://www.youtube.com/watch?v=EllbltLLnzo

Stephen Hawking's Brief History of Time is a scientific book, one which is convincing. The Universe came into existence in a Big Bang from a primordial atom and took 13.82 billion years to make our galaxies and stars. If anybody could give us truth about our beginning, it is Stephen Hawking, one of the most brilliant thinkers of all time, right up there with Newton and Einstein. He published his Brief History in April 1988. Stephen Hawking had Amyotrophic Lateral Sclerosis (ALS), also known as Lou Gehrig's disease. He died in 2018 at the age of 76. Why did the universe spring into life and will it ever come to an end? These were large questions which he wanted to answer before he died.

Roger Penrose gave Hawking an idea. If a big hole in space can make objects disappear, like a black hole, then couldn't objects also come out of nothing? Hawking likes Einstein's idea that space and time should be looked at one, as space-time, and that gravity, depending upon the object's size is suspended by the curve of spacetime. Hawking realizes that this bend in space between objects is not a force but the dips that pull objects toward each other. What we experience as gravity is actually the curve of space-time around us.

The hole in space is created when a star collapses in on itself. A singularity is like a drain, a hole in which things fall down and disappear. Therefore, his theory says, all things sprang from a singularity. A reverse black hole showers everything out in a Big Bang. The universe suddenly appeared from nothing! The universe had a beginning, says Hawking, and "this was finally proved by Penrose and me." This was also the proof of the beginning of Time.

However, to understand how the universe can explode out of a tiny hole in space, Hawking needs to understand the theory of the very small, which is quantum mechanics. The problem is that the theory of general relativity and quantum mechanics are incompatible. If we zoom into the quantum world, the laws of general relativity don't apply. We now know that electrons zip around the nucleus in an amorphous cloud and can even be in several places at the same time. To combine these two theories into one is to combine them into a Theory of Everything. Nobody's gotten to that solution yet!

Waves in space create particles. Every particle has an anti-particle. They appear but then collide and destroy each other. No one knew how quantum mechanics played on the larger scale of the universe. Hawking asked, what if I put a virtual pair next to a black hole? What happens?

Hawking realized that the negative particle is pulled into the black hole and the positive particle escapes as radiation. Hawking proves that black holes give off heat, called "Hawking radiation" in his honor. It's the closest anyone has ever come to a Theory of Everything.

Hawking discovers that the positive electrons which escape the black hole make the black hole evaporate like a puddle in the sun. As the black hole evaporates, it shrinks faster and faster until it becomes so small that it explodes in a Big Bang…recreating the universe. Could anything come out of the black hole when it explodes? This tremendous explosion from a tiny point. The Big Bang resembles a black hole explosion on a vastly larger scale. An understanding of how black holes create particles will lead to a similar understanding of how the big bang created everything in the universe. At the moment, nobody knows if black holes explode. Physicists are looking for gamma ray explosions which could prove that a microscopic black hole can explode.

There is a question: Why did gravity weaken just enough to stop the universe from collapsing on itself. We have to have a theory to explain gravity on a small scale, says Hawking. Gravity is much weaker than the other three forces, namely, weak nuclear, strong nuclear and electromagnetism.

Scientists thought that at the time of the Big Bang, electromagnetism and gravity were the same strength. As the universe expands gravity weakens. Nobody knows why.

Transcribing Hawking's ideas into the written word must have been a painstaking process for him, since his enunciation to his scribe was difficult to understand. And with pneumonia in 1985, his vocal cords were severed. A machine linked to a computer was invented to accommodate what Hawking wanted to say, so he could continue with his research. Hawking's "electronic voice" was born!

Parallel to Hawking's work is string theory which says everything was made of energy strings, which is one trillionth, trillionth the size of an atom. String theorists believe there are 9 extra dimensions through which gravity must expand. Gravity weakening is what stops the universe from collapsing back in on itself. String theory predicts we are surrounded by tiny extra dimensions. These extra dimensions trap gravity and weaken it.

Stephen Hawking put his faith in tiny black holes being created in the CERN collider which would prove string theory and the dissipation of tiny black holes. However, delays in upgrading the Collider have thrown a monkey wrench in Hawking's plan to live long enough to prove how string theory and black holes are connected.

Google: Amo Penzias and Robert Wilson detected the Microwave Background Radiation from the Big Bang in 1965, the clear proof of a Big Bang uniformly spread out in the Cosmos. Scientists can trace the Big Bang back to at least 10-43 seconds after the event, which is an incredibly short time. The earliest moment scientists discuss is $t = 1 \times 10^{-43}$ seconds, which is the number 1.0 with the decimal point moved 43 places to the left. It's amazing how all atoms came out of hydrogen and helium with different combinations of their isotopes and with heavy elements out of exploding stars. Each galaxy moving away from each other has different times compared to us. Isn't that fascinating? We are stardust.

Until roughly 380,000 years after the Big Bang, the entire universe was a thick opaque cloud of plasma of electrons and nuclei freely floating around. As the universe expanded, it cooled off enough to let the plasma become atoms, the first ones being hydrogen and helium. At this time, the universe became transparent, so that you had light. We observe the light from this time as the cosmic microwave background (CMB).

Roger Penrose sees this construct of the universe's evolution as repeating in a cyclical pattern forward and backward in time into eternity, so there was no real beginning and no need for God.

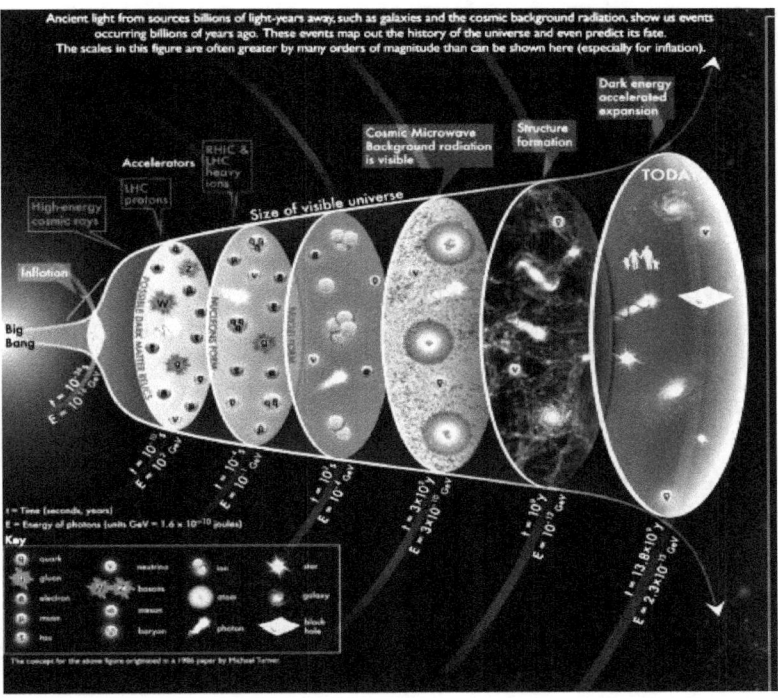

Perspective

Our brain gives us our perspective on life and the universe in which we live. We have opinions about God, death, the afterlife, and our own self-awareness. The brain is where our understanding is. The philosopher, Rene Descartes, asserted: Cogito, ergo sum! I think; therefore, I am.

If we did not exist, our understanding would not exist and if there were no one in existence to understand the universe, would the universe therefore really exist? Maybe the better question would be, without us, would the universe have any meaning?

When I studied Physics in grade 11, I was amazed at how the atom had electrons in orbit around the nucleus. I thought of the sun's solar system. We are so tiny in the grand scheme of things. Our planets orbit our suns like electrons around their nuclei.

Years later, with the magic of the computer, I read about a study carried out by astrophysicist Franco Vasa and neurosurgeon Alberto Feletti. They saw a similarity between the brain's intricate connections of neurons and the universe's cosmic web. The patterns they form are strikingly similar. Some researchers have boldly claimed that the universe itself may think or possess a form of consciousness. They see the brain as a 3-pound universe! The human brain's neuron Network and the cosmic web of galaxies have a striking resemblance.

This leads to pan-psychism, a theory that everything has a mind-like quality. Italian philosopher, Franchesco

Patrizi, coined the term pan-psychism in the late 16th century. Even Plato in ancient Greece said that the world was a living being endowed with intelligence. Developments in neuroscience, psychology and quantum physics have contributed to this renewed interest in pan-psychism which suggests that Consciousness is a fundamental part of reality and that all matter, even particles, have some form of basic consciousness. Studies now suggest that consciousness manifests itself through quantum fields. Cosmic psychism suggests that the universe itself is conscious. Yet, skeptics have doubts. Just because two things look alike does not mean they function the same way. Some scientists wonder if humans have a tendency to put their own attributes into what they see. That tendency is called anthropomorphism. For instance, in the industrial revolution, we saw the brain as a machine; in the digital Revolution, the brain was compared to a computer.

We now find ourselves in the context of artificial intelligence and machine learning, so that now we have a new metaphor where we see the universe as a vast neural network, i.e. the cosmic web! It is no wonder that we see the cosmic web as having the same make-up as our brain of 100 billion neural connections which form nodes, filaments and clusters that shape thoughts and feelings. These neural networks make up 25% of your brain with the rest as 75% water.

In a bizarre coincidence, our observable universe has 100 billion galaxies. Scientists estimate that 25% of the matter in the universe is visible, whereas the other 75% is dark matter. The universe's cosmic web and the brain's web of neurons, therefore, look alike! This, of course, is the basis for researchers wondering if the cosmos is indeed a vast cosmic brain.

*****

Excerpt from the Transcript:
https://www.youtube.com/watch?v=suuA3QGUPis&t=6s
Matter is no longer considered the foundation of life; instead, it is part of a Continuum involving various evolutionary processes connected to the quantum field since everything in the universe ultimately arises from quantum mechanical events.

It is believed that the universe is governed by a greater consciousness, an overall field overseeing its intricate workings. Understanding Consciousness may require an analytical study of higher States of Consciousness forming the basis for new theories and experiments through studies of the Quantum Mind and the exploration of quantum realities. Living beings are no longer viewed merely as chemical generators; instead, they are seen as part of an energy Network, a collective field that moves us beyond our separate selves into a unified structure of the universe.

This vastness allows us to direct our minds, hearts and feelings toward the great mystery and wonder of universal existence experiencing the profound reality and Majesty of the cosmic domains of life. Before us lies The Mystery of Life and Consciousness with the exciting possibility of exploring the universe within each of us.

The human journey in the information age has just begun if we approach it with love, discipline, passion, curiosity and courage.

We can use our greatest tool to explore the world of ideas and the true meaning of the universe. This incredible tool is called Consciousness. It's our birthright! Perhaps,

Consciousness didn't simply emerge or flicker into existence. It has always been present, inherent in us and everything around us. The journey towards understanding reality and Consciousness is an Ever-evolving Quest, a continuous exploration that invites us to challenge our assumptions, expand our Horizons and embrace the Limitless possibilities that lie in the vast expanse of the unknown.

God or Nothing?

Stephen Hawking view on God | Science vs God
https://www.youtube.com/watch?v=EVv17yD3KUI

Astrophysicist, Stephen Hawking, explained the universe's evolution in A Brief History of Time, a brilliantly written short book on the Big Bang and how the universe evolved since then 13.8 billion years ago. I enjoyed reading it.

Stephen Hawking claims that mankind has used God to fill the space for unknown answers. That concept of not knowing is called the God of the Gaps. What mankind couldn't understand was categorized as coming from an entity called God. There was no explanation of how water came to be, what made the sun and moon shine and how the world was formed, so primitive man concluded that God created everything.

Before the Big Bang, there was nothing, zero, the positive and negative energy split apart at the time of the Big Bang. All the positive energies turned into atoms and became planets, stars and galaxies. All the negative energies were bound in space as dark energy. Only at the Big Bang did time come into existence.

According to Stephen Hawking, God would not have had time to create the Big Bang because time did not exist. Hawking says that the search for God is meaningless because there was no such thing as time before the Big Bang. Therefore, the universe was not created by anyone; it was created spontaneously. Question: If the universe came out of nothing, then who created nothing? Maybe that is a question that need not be asked, if the universe is a phenomenon that repeatedly creates itself.

*****

In contrast, I believe in the <u>God of the Needs</u>. That comes from human weakness, to which I admit. If you are sunk into depression, then you reach out for someone or something out there in the universe which will help you through a rough time. This is the God of the Needs. Mankind has built places of worship in recognition of such a need. It is comforting to rely on a power that exists outside of oneself so that one has the strength to survive through the toughest of times.

Yet, a believer in God need not hate science, and a science lover need not hate God.

Some scientists, even clever writers like C.S. Lewis, have conjectured that spiritual beings might exist who were capable of creating worlds and other realities. The Kardashev Scale also includes civilizations in existence which might be able to create galaxies and worlds.

This brings one into conjectures about super beings. The English poet, John Milton, wrote <u>Paradise Lost</u> about the war between spiritual beings, one group led by Satan and another led by Michael and those on God's side. After mankind was created, there were giants on Earth who were conceived when the Nephilim found the females of the Earth attractive. If giant dinosaurs roamed the Earth eons ago, then why not giant Nephilim? Another conjecture is that the Anunnaki from the planet Nibiru flew to Earth to genetically engineer the human race. It's amazing how myth and stories from the Bible interact about God creating the human species in ancient days?

The ultimate question about creation is: If God created the universe, then who created God? Again, that is an irrelevant question. The universe always was and always will be!

This comes from Roger Penrose's thesis, that the universe creates itself in a repeating cycle of Big Bang, Big Crunch and Big Bang. Who needs God then, if this is true?

My own belief does incline towards a universe coming out of nothing or a universe always being there. Nor do I subscribe to humans being genetically engineered by a race of interstellar super beings called the Anunnaki coming to Earth 500,000 years ago from another planet called Nibiru. Some scholars claim that these beings were giants and had superpowers. They gave ancient Sumerians the skills of writing and building. They ruled the Middle East thousands of years ago as powerful giants who purported to be gods. Goliath apparently was part of that race called Nephilim.

I also do not subscribe to Sir Roger Penrose's idea of a cyclical and self-replicating universe.

I'm content with my Faith resting in Jesus and a fatherly God who exists outside of time and space and is able to do what he wants with atoms and nothingness.

The Catholic Church was pleased when their astrophysicist, the Belgian priest, Father Georges Lemaître, worked out the mathematics for Creation proving to the Church that God created the universe out of nothing in a Big Bang. The priest was known as the father of the Big Bang theory through his Ph.D. proposal way back in 1927. It's amusing that Einstein had nothing to say to the priest about the mathematical part of his paper. Technically it was perfect, but he disagreed with the priest on the physics: "from the point of view of Physics this seems to me abominable." Well, Einstein was wrong, and Lemaitre was right, at least in the concept of a Big Bang, not in a proof that God caused it.

The biologist, Richard Dawkins thinks that people who believe in God are crazy; they are in a mass delusion about something that does not exist. But that's his opinion in the long run because you can't prove a negative. You can't prove that God does not exist.

To believers, that's alright because as Paul Tillich, the theologian says, Doubt is actually a part of the definition of Faith. This idea is referenced in Mark 9:23-24: "Jesus said unto him, if thou canst believe, all things are possible to him that believeth. And straightway the father of the child cried out, and said with tears, Lord, I believe; help thou my unbelief."

To many scientists, that God exists or does not exist, does not really matter. There is so much outside of God without bringing Him into the equation, which makes the universe a marvel in itself. String theory, nuclear forces, quantum field theory and black holes offer enough marvelous areas of research and discovery which keep scientists enthralled with the wonder of it all, with what a marvelous universe we have, even without God. Though, in my opinion, it would be a nice concession to include Him.

How Big is Big?

How Large is the Universe? Bigger than you can Imagine?
https://www.youtube.com/watch?v=m2YJ7aR25P0&t=68s

We still have a long way to go before our sun expands and swallows up the earth, estimates are 4 billion years. We only started space exploration recently. We are babies in space exploration and in understanding the universe. Only 12 people have walked on the moon, our nearest celestial body at 238,000 miles away from us. Mars is not far off at 140 million miles. There are so many things to discover both big and small.

The average distance between the earth and sun is 93 million miles. This is what we use to measure distance within our solar system, called an Astronomical Unit. How long would it take to get to the sun with a commercial airliner? Just 19 years by plane! Beyond the solar system, we use light-year distances, the distance that light travels within a year, which is quite fast and quite far! Light travels 186,000 miles per second in a vacuum. This is the speed limit of the universe. Nothing can travel faster than light! Albert Einstein made a breakthrough when he said that mass and energy were interchangeable; therefore, $E = mc^2$ -- where E is energy, m is mass and c^2 is the speed of light squared.

The closest sun to us, other than our own, is Proxima Centauri which is 4.2 ly away or 25 trillion miles. To get there by plane, it would take 5 million years and to get there by car, it would take 47 million years!

Our own galaxy, the Milky Way, is 120,000 light years across and 100,000 ly thick. Within this system, other stars have planets too, some of them within the Goldilocks zone or the habitable zone.

Our nearest other galaxy is Andromeda which is 2.5 ly away. Zoom out further and you see the Virgo Supercluster which has 20,000 galaxies. Zoom out more, and you get the Laniakea Supercluster which has 100,000 galaxies.

As we view the Cosmic Background Radiation and entire observable universe, we have to realize it's 93 billion ly [light years] across. Scientists estimate that the entire universe is at least 250 times larger than that, what we consider the observable universe. Science knows a lot but ultimately, we know nothing. We don't have all the answers!

Finite versus Infinite

Unravelling the Universe: Professor Brian Cox Speaker YouTube Series:
https://www.youtube.com/watch?v=PlvrzL_MWIg

Brian Cox is one of my favourite astrophysicists. He does not believe in God but his approach in explaining the universe is sensible and if I do say so, in a gentle and gentlemanly manner. During one of his lecture series, he quoted author John Updike who said, "Astronomy is what we have now instead of theology. The terrors are less but the comforts are nil." He points out that the number of galaxies in the observable part of the universe is estimated to be around 2 trillion galaxies. That is big!

Brian Cox's hero is Richard Feinman who marveled that our universe has atoms which actually contemplate atoms, that people as a conglomerate of atoms have brains that can think, "atoms with consciousness."

Cox explains that the Earth formed 4.5 billion years ago with the first half a billion years being devoid of life, called the Hedan Epoch. There was a dead world. Soon after the earth cooled and the oceans formed, around 4 billion years or so, life began, transitioning from geochemistry to biochemistry. Some scientists argue that once the oceans formed, the inevitability of life ensued, which took a long time, 3.5 billion years to only 500 million years when complex life evolved. Stromatolites became detectable as fossils. Colonies of photosynthetic bacteria. Human beings arise about 1/3 of the age of the universe, in the Rift Valley, about 200,000 years ago.

Brian Cox admits that we, humans, might be the only civilization in the whole of the Milky Way. We are the "Rare Earth Hypothesis". Therefore, we are precious and valuable as a species. On that basis, Brian Cox argues, "if we are the only civilization in our galaxy, then I think we may behave a little differently than the way we are behaving today." Brian Cox argues not only for evolution but for evolution that may be so rare that we would be stupid to destroy it. He touches on Einstein's Theory of Relativity 1905 and General Relativity 1915. Einstein in those days argued for a static universe, He did not like the notion of infinity. Whereas a colleague, the priest Georges Lemaitre, argued for an expanding universe that came from a primordial atom in a Big Bang. It took Einstein's introduction to Edwin Hubble who let Einstein look through the telescope on the Mount Wilson Observatory in California in 1931 which showed Einstein that the universe is ever expanding and not static.

In later years, Einstein considered his cosmological constant as a blunder to make his equations show that the universe was static when it wasn't. It wasn't until the 1960s when the Cosmic Microwave Background showed the heat signature of the entire universe when it first exploded in a Big Bang.

This gave us the age of the universe at 13.8 billion years. The furthest we can see is 380,000 years back when the first light was created. Protons and atoms could not form before that because the universe was too hot and dense to emit any light; the universe was opaque. At that point in the past, there were no stars, no planets and no galaxies. The areas of slight variations in the CMB eventually formed stars and galaxies. The way that galaxies expand in the theory of inflation indicates that dark matter exists which forces the universe to expand.

Cox refers to Richard Feynman's article on *The Value of Science* which stresses that the value of science is "not knowing" which is the driving force to discovery.

"What can we do to dispel the mystery of existence? By admitting that we do not know, we have found the open channel. Doubt and discussion were essential to progress and the unknown."

Feynman calls science the satisfactory philosophy of ignorance where you admit you don't know something and humbly pursue an answer to fill in what you don't know. "Doubt is not to be feared but welcomed and discussed and to demand this freedom as our duty to all coming generations."

*****

Is the Universe Infinite?
https://www.youtube.com/watch?v=dG1JpC5jels

How far do the stars stretch out into space? What's beyond them? Is there a border? If the universe is expanding; what's it expanding into?

Computer modeling has reconstructed much of what went on in the creation of galaxies and the expansion of the universe. Often they have to deal with the concept of infinity. The Greek, Zeno, imagined a situation where you are crossing a road, and you are faced with subdividing your progress by increments of halves. It could go on forever! A century later, Aristotle looked at infinity as the formless chaos from which the world emerged: a primordial state with no natural laws. Aristotle reserved the term, actual infinite, for the prime mover that created the world. This became the basis for the First Cause or the Cosmological argument for God.

Computer scientists recently calculated Pi to 5 trillion digits. Pi is the relationship of the diameter of a circle to its circumference. The largest number ever used is Graham's number which is a calculation of angles in a type of hypercube, larger than quinto-quadragintillion from the Buddhist Calendar. The ultimate ceiling is infinity! Is the universe infinite?

Muslim astronomers looked at the universe with advanced eyes. Only later, western astronomers like Tycho Brahe, Bruno and Galileo approached the investigation of stars and planets with advanced techniques, Galileo for example with his telescope on Jupiter. The Catholic Church, however, cut short this progress with the Inquisition and punishments like burning at the stake.

Einstein luckily came along at a time when the Inquisition was disbanded. Edwin Hubble and Milt Humason used the 100-inch telescope on Mount Wilson to confirm that the universe was expanding and not static. Einstein had to see this with his own eyes. Hubble's Law said that the further you looked, the faster galaxies were moving away.

Running the expansion backwards in time, astronomers judged that the universe started off with a Big Bang, suggested by the priest Georges Lemaitre, a Big Bang which happened 13.8 billion years ago.

Possibility of Multiverses?

Energy is constantly welling up from the vacuum of space in the form of particles of opposite charge, matter and anti-matter. In primordial times, energy fields embedded in this quantum vacuum suddenly moved into a higher energy state causing space and time to literally inflate. Our universe then burst forth in a big Bang. Astrophysicist Alan Guth suggested that inflation happened incomprehensively quickly. In fact, he suggested, our universe would have expanded like a bubble and joined other bubbles in a froth, expanding multiverses across an endless ocean of time and space. For the moment, this question whether the universe is finite, or infinite will always elude us. At the moment, we don't have the intelligence nor the equipment to find the answer to what lies beyond the observable horizon!

Part Two

Part Two	
Topics Discussed	
Being Honest with Yourself	p. 40
Carl Sagan's Cosmos	p. 41
The Big Picture: Sean Carroll	p. 43
Beyond the Observable Un.	p. 46
The Cosmic Scale	p. 47
The Quantum World	p. 52
Forces and Particles	p. 53
History of Quantum Mech.	p. 61
The Question of God	p. 66
Lawrence Krauss	p. 68
Richard Dawkins	p. 70
Roger Penrose: Cyclical	p. 71
Max Planck: Divine Mind?	p. 74
What About Aliens?	p. 78
The Kardashev Scale	p. 79
Proving God Exists	p. 84
Who Created God?	p. 86
Is the Universe Alive?	p. 91
Artificial Intelligence	p. 94
Yuval Harari	p. 98
Leo Tolstoy	p. 101
More YouTube Videos	p. 103
Conclusion	p. 113
Extra Thoughts	p. 114
John's Publications	p. 119
John's Photobooks	p. 125
Contact	p. 127

BEING HONEST WITH YOURSELF

Professor Brian Cox is my hero like Richard Feynman and Carl Sagan were heroes to him. He confesses that when he does not understand something he will admit it. To be honest, with yourself, you can then go after understanding.

He has transitioned through 3 careers: as a rock musician, as a physicist and as a TV lecturer. He's always been honest to admit that he does not know something when he does not. That, he feels is the beginning of science to learn what you do not know.

Professor Brian Cox: How To Find Your Place in The Universe
https://www.youtube.com/watch?v=lsCC_G5G9YE
People who do not understand something fall back on jargon and they wave their hands trying to hide that they are flustered. The reason Brian Cox likes Feynman is his honesty. Feynman called science the satisfactory philosophy of ignorance. "You have to be delighted to not know." Progress then comes from admitting ignorance and then trying to find out answers with a passion just to know. You have to jettison the fear of the unknown. Brian Cox says a value should be placed on acquiring knowledge and not on whether you are right!

He also sees a responsibility adhering to the things you learn, to pass them on with honesty. Brian Cox is a life-long learner. He admits that he is slow at learning something and that he needs patience for things to sink in. Patience is a luxury in the modern era. He finds himself fortunate. He finds it astonishing that we exist, that we live in a beautiful universe, that we need to notice that something is worthwhile and that we are lucky to live in a time where you can pursue music, the arts and higher pursuits of knowledge, wherever your passion is. Brian Cox references Richard Feynman again from his pamphlet, The Value of Science: "What we have to learn is how doubt is not to be feared but welcomed and discussed." His message is that if people don't get it, then you should learn to fill doubts with knowledge. In fact, the theologian Paul Tillich wrote a book <u>Dynamics of Faith</u> which explores the idea of faith and the role of doubt in it

Brian Cox noted that we are probably the only civilization in our Milky Way Galaxy, the only intelligent life. Then the question becomes profound, of what does it mean to be a human in the universe. Meaning is a property of intelligence, so meaning exists here. But if there is no other intelligence out there in our galaxy and if we destroy ourselves, then we might eliminate meaning in a galaxy of 400 billion stars forever! So, consider that, says Cox to world leaders! "You have a galactic size responsibility to maintain meaning in a galaxy. A lifeless world, a lifeless galaxy is a meaningless galaxy."

Carl Sagan's Cosmos - Episode 8 - Journeys in Space & Time

https://www.youtube.com/watch?v=abp3q7aYOss

But if we do not destroy ourselves, I believe that we will one day venture to the stars. When our solar system is explored, the planets of other stars will beckon.

Space travel and time travel are connected. To travel fast into space is to travel fast into the future. We travel into the future although slowly all the time but what about the past? Could we journey into Yesterday. Many physicists think that this is fundamentally impossible, that there is no way we could build a device which would carry us backwards into time. Some say that if we could build such a device, it wouldn't do us much good, that we couldn't significantly affect the past for example, suppose you traveled into the past and somehow or other prevented your own parents from meeting. When then, you probably would never have been born, which is something of a contradiction since you're clearly there. Other people think that the two alternative histories have equal validity that they are parallel threads of time, that they could exist side by side: the history where you were never born and the history that you know all about. Perhaps time itself has many potential Dimensions despite the fact that we are condemned to experience only one of those dimensions. But what about changing history in a major way let's say persuading Queen Isabella not to bankroll Christopher Columbus. Then you would have set in motion a different sequence of historical events which those people you left behind in our time would never get to know about. If that kind of time travel were possible, then every imaginable, then every sequence of alternative history might in some sense really exist. Would a time traveler be able to change the course of history in a major way? The ancient Greeks imagined history to be a complex multitude of deeply interwoven threads, biological, economic, and social forces that are not so easily unraveled. The ancient Greeks imagined the course of human events to be a kind of tapestry created by three goddesses, the Fates. Random, minor events generally have no long-range consequences but some which occur at critical junctures may alter the weave of history. There may be cases where profound changes can be made by relatively trivial adjustments. What if our time traveler had persuaded Queen Isabella that Columbus geography was wrong. Almost certainly some other European would have sailed West to the new world soon after. The discovery of America was inevitable. Of course, there wouldn't be any postage stamps showing Columbus and the Republic of Columbia would have some other name, but the big picture would have turned out to be more or less the same. It's a lovely fantasy to explore those other worlds that never were!

Carl Sagan talks about evolution in the last part of Episode 8. He says, "we are star stuff which has taken its Destiny into its own hands. The loom of time and space works the most astonishing transformations of matter. Our own planet is only a tiny part of the vast Cosmic tapestry, a starry fabric of Worlds yet untold. Those worlds in space are as countless as all the grains of sand and all the beaches of the Earth. Each of those worlds is as real as ours, and on each of them, there's a succession of incidents, events and occurrences which influences their future. Countless worlds, numberless moments, an immensity of space and time and on our planet, we face a critical branch point in history…it is well within our power to destroy our civilization and perhaps our species as well…if we capitulate to stupidity we can plunge our world into a Darkness deeper than the time between the collapse of civilization the Italian Renaissance…but we are also capable of making a meaningful life for every inhabitant of this planet to enhance our understanding of the universe and to carry us to the stars. "

The Big Picture: From the Big Bang to the Meaning of Life - with Sean Carroll
https://www.youtube.com/watch?v=2JsKwyRFiYY
Sean Carroll demonstrated to his audience that he moved a book on a desk by putting pressure on it with his hand. The thing is moving because this other thing is pushing it.

If you stop pushing the book, it returns to its natural state. "If you trace the chain of motion and movers backward you eventually would need to reach an unmoved mover and thereby prove the existence of God!" Everything that happens, happens for a reason. This is the principle of sufficient reason. Aristotle, Spinoza and Leibnitz were of the same mind, so that we can understand how the world works at a deep level by providing explanations for everything we see. Nothing happens randomly, nothing just happens. There is always a purpose, a cause a reason why things happen?

Sean Carroll asserts that the world does not work that way at a fundamental level. Bertrand Russell apparently tried to fix how we view causality: "the law of causality is a relic of a bygone age." Cause and effect is no longer fundamental to our understanding of how reality works. Carroll asserts that cause and effect is not a fundamental principle. It's a useful tool in the macroscopic world where you see cause and effect, but it is nowhere to be found in the most fundamental laws of physics within quantum physics.

Sean Carroll says we understand reality on two levels. The Macroscopic world considers... cause and effect, reasons why, dissipation and the arrow of time. Microscopic fundamental physics considers... laws of nature, patterns, differential equations and conservation of information. Different levels of description involve completely different vocabularies.

With regard to the arrow of time, entropy was low in the past, 13.8 billion years ago and has been increasing ever since. Entropy is the second law of thermodynamics! It progresses from order to disorder with time. It starts out with simplicity and evolves to complexity at which point life and humans came to be. It's ironic that entropy increases with time towards more disorder because nature has a complex reaction to entropy which is the passage of time and the rise of life! With the rise of life, you have the rise of evolution! With evolution you have the human species, you have imagination and consciousness.

Doctor Carroll comments that we don't know how consciousness evolved. But it's there, and we have free will and the ability to choose right from wrong. He says that if we look at everything as atoms, we don't have a choice, but we are more than that. Everything as atoms is determinism. Rene Descartes told Princess Elisabeth that the soul interacts with the body via a body/mind dualism. He said, "I cannot doubt the existence of my mind because it is the thing doing the doubting." He came up with the argument, "Cogito, ergo sum." I think, therefore I am! However, he asserted that it was easy to doubt the existence of his body, therefore there must be two separate things.

Naturalism says that there is only one world, the natural world. Poetic naturalism says that there's many ways of talking about the world. In other words, one must evoke an emergent vocabulary which is able to handle new ideas. For example, you can describe yourself as a person, you have choices. If you describe yourself as a conglomerate of atoms, you don't have choices. Carroll suggests that life is better as a person with the emergent property of a being who can make choices.

I feel uncomfortable with Professor Carroll's last part of his lecture. He suggests there is no absolute to right and wrong. A person can decide what is right and what is wrong. You construct your own morality. Right and wrong seem to be a consensus among communities of people. Morality is made up like the rules of chess. When we make up the rules to right and wrong and living together we have goals in mind.

It is interesting that he wraps up this sidenote on morality by saying, "It doesn't mean that just because the universe doesn't tell us how to behave, there is no way to behave." I'm sure that an argument can be made for morality to be more than subjective, that it may not be a relative thing that we construct and choose as life goes on. I prefer to choose that there are absolutes in the choice of morality but for now, let Doctor Carroll have his say.

Doctor Sean Carroll ends his lecture with an acknowledgement that we are all going to die someday. We are finite. With that in mind, "we are the little part of the universe in this age when things are complex and interesting, where we have developed the capacity for self-awareness and reflection and thinking and rational thought and writing books."

"We have the ability to be rational to think to invent to discover to create new things to care about each other in ways that other parts of the universe just don't care…we have a choice."

Beyond the Observable Universe
https://www.youtube.com/watch?v=mty0srmLhTk

The Size of the Universe could be finite or infinite. Either way it is huge! Its age is 13.8 billion years. So how can the Hubble Telescope see the light of GNz11 a galaxy over 33 billion light years away, which is almost three times the age of the universe away? The explanation lies in the expansion of time/space, that our galaxies are located on the edge of what you might consider a balloon, and it is this balloon which is expanding faster than the speed of light. It is the edge of our universe that is ever expanding its boundaries. This leads us to another question. What is the shape of the universe? It could be flat, saddle shaped or doughnut shaped, toroidal shaped and funnel shaped.

For more information about the Cosmic Microwave Background radiation, dark energy and the multiverse theory, check this link on How Large is the Universe? Like bubbles, they are rising up continuously in the oceans of infinity.

In the grand scheme of things, we have shrunk in importance since Galileo's time, as we discover bigger things in the universe. I am reminded of Jim Carey's version of "The Grinch Who Stole Christmas", where the "Whos" live on a snowflake as the camera pans outwards at the end of the movie from their tiny world.
https://www.youtube.com/watch?v=yaX4iGw-b_Y

The Cosmic Scale
https://www.youtube.com/watch?v=4iC9Qi3y9q8
Our galaxy, The Milky Way, contains some 400 billion stars, and many of them have their own solar systems, with possibly habitable planets. When you consider other galaxies in our neighbourhood, one asks, how large is the universe itself?

The observable universe is around 93 billion lightyears in diameter. This is in contradiction to what we consider the age of the universe as being only 13.8 billion years old. The difference is accounted for between the first Big Bang and the rapid expansion of space where our galaxies are flying away from us faster and faster. Edwin Hubble noticed the difference between red shirt and blue shift, with the red shift phenomenon stretching space away from us at unimaginable rates of expansion. Scientists have estimated that 2 trillion galaxies are located within the observable universe and one septillion stars within the known universe! And yet that makes up only a small percentage of the universe's total scale, which could be millions of times more voluminous than we can count.

So, what is the universe? The universe is defined as all of space and time...all points in space at all of its life. Space is the volume within the universe, the three-dimensional field which lies beyond Earth's atmosphere, acting as the vast backdrop to the universe's matter. Time is the constant flow of cause to effect within this space, the catalyst of life, death, creation and evolution. Space and time are fundamentally linked, and they combine to form a system of nearly unknowable complexity known as "space-time", which is often referred to as four-dimensional. Space, therefore, constitutes 3 dimensions, and time is the 4th dimension and any cube cross section along this axis represents the universe at a certain point in its life. The understanding of "spacetime" goes back to Albert Einstein.

Mass is a form of energy, which has a warping effect on the fabric of space. We see this as the effect of gravity.

The YouTube video, "The Cosmic Scale", https://www.youtube.com/watch?v=4iC9Qi3y9q8, does a superb job of explaining the concepts which come out of Einstein's theories. Up to 1929, at the time of Edwin Hubble's study of galaxies, it was believed that our galaxy was the universe and that it was static. Even Einstein had believed in a static universe but changed his mind when Hubble had Einstein look through his telescope because by 1929, Hubble had discovered that we are not the only galaxy in the universe.

If our Milky Way Galaxy is large, can you imagine the size of all other galaxies in our universe?

Just to get a handle on the size of our galaxy, The Milky Way, Google says it is approximately 100,000 light-years in diameter. Our solar system is 26,000 light-years from the center of the Galaxy. All objects in the Galaxy revolve around the Galaxy's center. It takes <u>250 million years</u> for our Sun (and the Earth with it) to make one revolution around the center of the Milky Way.

The Cosmic Scale video synopsizes the events of the Big Bang Theory as they happened almost 14 billion years ago. The Hubble Space Telescope launched in 1990 allows us to see a much younger universe. The farther the telescope can see the farther back in time it sees. The Hubble Deep Field image snapped a photograph from the constellation of Fornax. To see further back in time than this, images must be taken through infrared radiation. This ability came through the James Webb Telescope launched into a million miles from earth on Dec. 25, 2021. The JWST is able to see almost the whole way back to the beginning of the universe around 13.7 billion years ago. Redshift is a phenomenon which tells us that a galaxy is moving away from us. Blueshift is the opposite.

The shape of the universe is anybody's guess. It could be a globe, flat or doughnut shaped. Physicists at Oxford University estimate that the total area of the universe must be at least 250 times the radius of the observable universe. Now that is huge! This puts the total diameter of the universe at over 23 trillion light years. That means if our observable universe is home to 2 trillion galaxies, then the <u>unobservable</u> universe may house as many as 30 quintillion of them! With those numbers, it is estimated that life must be playing out on another world.

Reason would dictate that at some point the universe would slow down its expansion because the huge gravity from huge masses would attract...but that is not so. The existence of dark energy did the opposite effect, to increase expansion and not attract it. Dark energy is the invisible force which accounts for the expansion of the universe. We know barely anything about it. It's like an anti-gravity force. Dark energy does not interact with light so it cannot be seen or detected. It is suggested that dark energy comes from nothing and that it appeared out of nowhere about 5 billion years ago. As the universe expands, so does dark energy. Dark energy makes up about 68% of the universe, which is a heck of a lot of nothing! We cannot see dark energy.

Astrophysicists predict that the universe will expand so fast and large, that it will reach a heat death. It will become a boundless void and be dominated by dark energy. The universe will end in an age of darkness!

*****

Unlike normal matter, dark matter does not interact with light nor the electromagnetic force. This means it does not absorb, reflect or emit light, making it extremely hard to spot. In fact, researchers have been able to infer the existence of dark matter only from the gravitational effect it seems to have on visible matter.

Dark matter makes up about 27% of the universe. Dark matter is responsible for the structure of galaxies.

Dark energy and dark matter make up almost 95% of the universe. Dark energy is responsible for the accelerated expansion of the universe.

Ordinary matter, which is our observable universe, makes up ONLY 5% of the universe!

THE QUANTUM WORLD

With the large hadron collider and electron microscopes, we are getting closer to taking images of the tiniest things. In fact, Scientists claim they can take an X-ray picture of a single atom. They also say they can take an image of the shape of a light particle, a single photon which is a quantum breakthrough.

This breakthrough will give us advanced tools in medicine and research. Harnessing the power of quantum mechanics will give us quantum computers.

How Did We Go from NOTHING to EVERYTHING
https://www.youtube.com/watch?v=1LATf9AwORg
Our universe [allegedly] emerged from a place where there was no space, no time and no matter! Quantum Mechanics applies to particle physics, the smallest bits of existence. In that realm, rules seem to be thrown out the window. Quarks, electrons and neutrinos behave in strange ways, challenging reality and time. There are 4 fundamental forces:

1. The electromagnetic Force: powering electricity and magnets.
2. The weak nuclear Force: responsible for radioactive decay
3. The strong nuclear Force: which holds quarks together inside atoms.
4. Gravity: which attracts bodies together.

The big questions, however, are still not answered. What is everything made of? And how did everything come from nothing! Inventions have come from research into these questions. For example: the World Wide Web was invented at CERN, where the Large Hadron Collider is, which helps scientists share data more easily.

Space-Time: The Biggest Problem in Physics
https://www.youtube.com/watch?v=RIqVnFtOSr4
What is the Planck length? 10^{-33} It is a trillion, trillion, times smaller than an atom. When we try to figure out what's happening at that length, the laws of nature break down. At the fundamental level of nature, do questions of where and when even have answers? Einstein suggested that we should think of space-time as a unified continuum. When we look at that idea, we conceive of a grid which curves called a "manifold". Einstein's insight is that space-time bends and curves in the presence of matter and energy. The effects of this curvature produce what we experience as gravity. At the Planck length if we try to put coordinates on our manifold they lose meaning.

The hadron collider can measure length at 10^{-17} cm Spacetime below the Planck length does not have any meaning. The holographic principle says that space-time is like a hologram projected from the information available on some lower dimensional surface like the boundary of the universe. Space-time is a difficult problem!

All Fundamental Forces and Particles Explained Simply | Elementary particles
https://www.youtube.com/watch?v=-L5OQp2J46g
Molecules are small. But smaller than that are atoms. There are huge gaps of space between atoms; picture it like the distance between planets. Inside our bodies, that are made up of trillions upon trillions of atoms is empty

space! Clouds of electrons zip and sizzle at the outer edge of the atom. In the center of the atom, we have the nucleus. The nucleus and the orbiting electrons are collectively called an atom! What is the nucleus made of? We see protons and neutrons sticking together. These are subatomic particles.

Inside the proton, we have 3 tiny dancing particles. These are quarks. They are elementary particles or fundamental particles. The standard model of elementary particles comprises 12 matter particles and 4 force carriers in this standard model.

It's more correct to say that quarks collectively form a proton, than a proton "contains" three quarks. Protons are not elementary particles; they are subatomic particles. The finest and best-known building blocks of our known universe are quarks!

Elementary particles are subdivided into Quarks, Leptons and Bosons. Quarks and Leptons together are called Fermions. All elementary particles have 3 basic properties: mass, spin and charge.

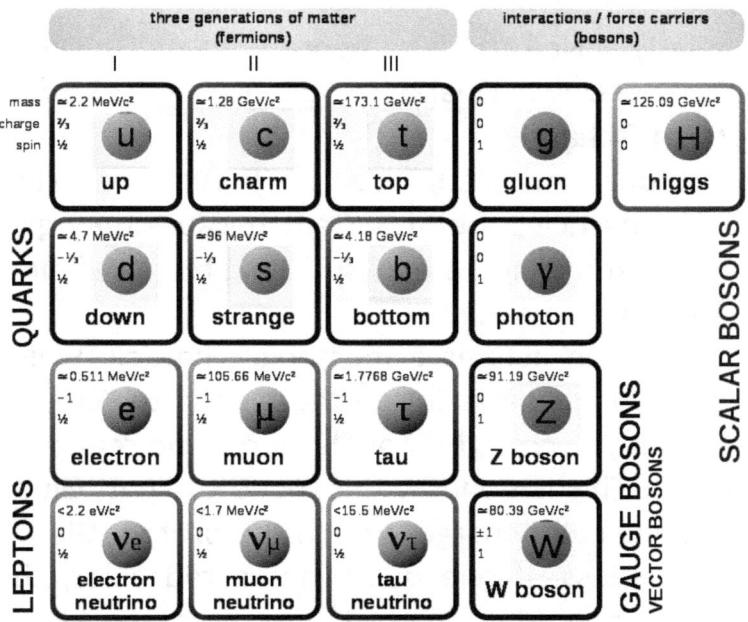

For example, the first quark has a mass of 2.4MeV/c2. This charge comes from $E = mc^2$.

Now the spin! Fermions have a ½ spin while bosons typically possess integer spins such as 0, 1, or 2. Spin does not mean the same as the spinning of a ball.

There are 3 generations of quarks. There are 6 flavors of quarks: Up, Down, Charm, Strange, Top and Bottom. The Up quark has a charge of +2/3, while the down quark carries a charge of -1/3. These charges apply to subsequent generations.

A Proton is composed of a charge of +1.

From two Up Quarks One Down Quark: +2/3 +2/3 − 1/3.

Similarly, a Neutron is composed of One Up Quark and Two Down Quarks: +2/3 − 1/3 − 1?3 for a total of 0 charge!

Protons and Neutrons are Baryon particles. Baryon particles are important because they give mass to atoms and are the building blocks of most matter in the universe.

The gluon acts as a messenger particle which allows the 3 quarks inside the proton to "change color", just a change, though there is no real color change. This is done at the speed of light. Gluon exchange color charges within the proton. Gluons are massless. If one quark has a red color, the other two quarks have to have a green and blue. Color is just a convenient name. 3 quarks in a baryon particle always have different colors by exchanging gluons.

The strong nuclear force binds the baryon particles together. Its range is 2.5 Fentometers, tiny, tiny, tiny.

A Meson is formed from a quark antiquark pair. Mesons decay quickly.

Leptons is another 6 pair: Electron, Neutrino, Muon, Muon Neutrino, Tau and Tau Neutrino.
The electron is very small. A Neutrino is even smaller. Neutrinos rarely interact with matter and

appear to go through the planet and humans with no problem.

Electrons orbit the nucleus like a cloud. They carry a negative charge. They buzz around the nucleus at different energy levels. When an electron jumps to a lower energy level, it emits a tiny packet of electromagnetic radiation – a photon. It has no mass and travels at the speed of light.

Virtual Photons cannot be detected. However, Photons can be detected as light.

In nature everything seeks a stable position. Atoms can become unstable. To seek stability, the boson is created to equalize the charge. There is also an antielectron charge. Nature has created particles and antiparticles. It's ironic because that might also apply to the macrocosm, a black hole having a possible white hole, though no one found one yet. The other name for an antielectron is positron.

This is basically where my understanding of all the interplay of particles got lost. The video talks about beta decay. Since it's close to the end of the video, I'll leave you floundering here to understand the rest of these changes and changes in the atom.

The weak force is responsible for nuclear fission. There's also a W boson and a Z boson which decays, producing another particle and antiparticle pair. The charge of the Z boson is 0.

The weak force range is very small.

The lifespan of these W and Z bosons is so short that they are only detectable in a particle accelerator like Large Hadron Collider in Switzerland.

The last item of importance in this video is the Higgs Boson which was detected in 2012. It is believed to be responsible for the mass of all elementary particles, incl. the W and Z bosons. Giving mass to all other elementary particles earned it the name the God particle.

If you lasted to the end of the summary of this video, you get a gold star for endurance. The cheque is in the mail, but don't wait too long!

Physicists Proved the Universe Doesn't Exist
https://www.youtube.com/watch?v=4hddxkhta3w
What is real? In the Heisenberg Uncertainty Principle, the observer effect can disturb the state of a quantum object where the observer cannot determine position and state of an object at the same time. The more precisely we can measure one property like position, the less we can know the other property like momentum. The quantum world does not follow our linear view of time. Instead, the quantum world looks at time in a fluid way where the past, present and future are one. That's about as spooky as quantum entanglement.

The Block Universe Theory encompasses both time and space. It looks at the Universe as a block where all things happen at all times at once. The concept is of a 4 block of space-time. All events from the origin of the universe to its end coexist simultaneously. Space-time can be thought of as a 4-dimensional block containing everything that has happened or will happen in the universe. All space-time encompasses all possible events.

Our limited view of time is a fabrication which is an illusion of how time works in a linear way. People, each lead their own world line. This asks a question about free will! If all possibilities exist within a 4-D construct, then is our free will determined? Is free will an illusion?

The speed of light is the cosmic speed limit, according to Einstein. But are there ways to break this speed limit and still not break the laws of Physics. Tachyons, for instance, are particles that can move faster than the speed of light, FTL. Tachyons have not been confirmed yet. However, the Alcubierre Drive suggests a warp drive where the space is compressed in the front and stretched in the back of a spaceship. The ship will move on this wave called a warp bubble. So far, this is only theoretical. Yet another theory suggests that the Einstein-Rosen-Bridges wormhole might cut through space with such tunnel constructions. Wormholes are not stable and have not yet been found.

However, information might be transmitted faster than light through quantum entanglement. Unfortunately, scientists have not found a way yet to control the entanglement of particles so that they can transmit information.

What if gravity is nothing but an illusion? Isaac Newton got the notion for the Universal law of gravitation by watching an apple fall. All objects with mass and energy are attracted to each other. Newton saw gravity as a force. Einstein's gravity model suggests that gravity is not a force, it is a curvature of the space/time fabric.

John Wheeler said,
"Spacetime tells matter how to move;
Matter tells spacetime how to curve."

Ordinary Matter = 5%
Dark Matter = 27%
Dark Energy = 70%

Einstein's Theory has problems of agreement between the large world and the small. A way to look at the difference is:

GenRelativity > Causality
Quantum Physics > Probability

THE ENTIRE HISTORY OF QUANTUM PHYSICS Explained in One Video
https://www.youtube.com/watch?v=4LychvQ-BAU
Neils Bohr and Werner Heisenberg collaborated from 1924 to 1927 to develop The Copenhagen Interpretation, a dominant understanding of quantum physics where both the wave function and the particles function had to be considered.

Heisenberg Uncertainty Principle states that it is impossible to know simultaneously a particle's exact position and momentum [i.e. the product of mass and velocity with perfect accuracy]. At the quantum level, the more precisely one property of a particle is measured, the more uncertain the other becomes. When we measure one property, information of the other is lost. Particles can exhibit both wave and particle properties, but it is impossible to observe both simultaneously.

According to the Copenhagen Interpretation, a particle transitions to a definite state only through observation of the wave function which collapses from a sea of possibilities to a single specific state. Particles can exhibit both wave and particle characteristics, but it is impossible to observe both simultaneously. When we measure one property, information about the other is lost. When we observe electrons creating interference patterns as waves we cannot determine their exact position. Conversely, when we measure electron specific position, we lose insight into their Wave Behaviour. According to Bohr, this is where classical physics at the quantum world does not apply. According to Bohr, the classical universe does not apply to the quantum world. The universe is probabilistic, not deterministic; it is interactive with the Observer, reality is shaped between the interaction of the Observer and the system. Measurement and the Observer play a central role in physical processes.

Albert Einstein did not accept this interpretation of reality. He called quantum entanglement a spooky action at a distance. Yet, two particles at the opposite ends of the universe could mirror each other, when you change one, you change the other.

Paul Dirac, an English physicist, proposed more complications into quantum physics. He said that every particle must have an anti-particle with the same mass and opposite electric charge. This was the prediction of antimatter, a form of matter not yet observed. Dirac named the opposite of the electron, a positron. This was substantiated in 1936 by American physicist Carl Anderson. Today this discovery of antimatter is used in positron emission tomography which uses gamma rays in the interaction between positrons and electrons to obtain detailed images of the internal structure of the human body.

Wolfgang Pauli's research discovered the Pauli Exclusion Principle where no two electrons within an atom can occupy the same Quantum State. They cannot have the same set of quantum numbers; EACH ENERGY LEVEL CAN ONLY ACCOMMODATE A CERTAIN NUMBER OF ELECTRONS DETERMINING THE STRUCTURE AND CHEMICAL PROPERTIES OF ATOMS. This showed why the periodic table has its current structure and also showed that electrons have a property called spin which only takes on two possible values. The periodic table was now arranged logically according to atomic numbers and electron configurations making it possible to predict the bonding tendencies of atoms. This opened the door to the discovery of fermions. The research of the small level also helped physicists to understand the behavior of large structures like white dwarf stars and neutron stars.

QUANTUM FIELD THEORY showed that particles and fields are actually inseparable from one another. Energy and matter are at a continuous interaction at the subatomic level. In the fabric of the universe, a field had to exist before a particle could emerge. The standard model is a framework showing how protons, neutrons, electrons and other particles came together. Dirac's mathematics reappeared in the research of Quantum Field Theory with the discovery of more anti-matter particles. The Large Hadron Collider was constructed to discover particles predicted by the Quantum field. We've discovered quarks, gluons and carriers of the weak Force. Through our understanding of the quantum field theory, we not only have an understanding of the small but also of the large, how galaxies, stars and planets were formed. We have a better understanding of the electromagnetic Force, the weak Nuclear Force, the strong nuclear Force, and Gravity. Discoveries in these areas have given us transistors and medical imaging devices based on Quantum Field Theory.

*****

The Latest Quantum Physics Breakthroughs II Quantum Space Documentary 2024
https://www.youtube.com/watch?v=qglkhd2EHIY

Scientists call quantum physics the hidden reality of the universe. Astrophysicists are fascinated by quantum entanglement and quantum superposition. There is also a study of how axions interact with pulsars. Axions are a strong contender for dark matter. Pulsars spin so rapidly that their magnetic field is a trillion times stronger than earth. The video comments that if axions are present, they can create unique spectral signals. Pulsars offer a one-of-a-kind setting to study dark matter. Axions and photons might interact producing detectable conversion signals in pulsar's radio waves.

The study of pulsars starts with neutron stars which exploded in a supervova. Superfluidity naturally occurs in neutron stars. Quasars were used to test the strange phenomenon of quantum entanglement. Quasars are powered by black holes. Quantum entanglement is a phenomenon where particles influence each other across vast distances. Einstein called this "spooky". Studies of Pulsars, Quasars and Neutron stars reveal how all things in the cosmos interact. Axions, superfluid states and entanglement are not just theoretical; they are real.

This has also opened new concepts of quantum tunnelling. Particles can pass through barriers that classical physics says should block them. Tunnelling actually creates the nuclear fusion that keeps the sun heated. Quantum rules have influenced phenomena like superfluidity, fusion and tunnelling. Quantum rules have influenced the formation of stars, galaxies and the matter around us. From the formation of electrons to the formation of molecules in space, scientists are bridging the gap between the quantum world of the small to the cosmic world of the universe. Indeed, Quantum rules govern phenomena across all scales.

Quantum tunnelling may play a role in the formation of strange matter. Experiments are also made into synthetic quantum dimensions. Another factor: quantum gravity at the Planck scale, which seeks to unify quantum gravity and general relativity. Classical physics fails to describe phenomena at the Planck scale where time and space are thought to become "quantized".

What is the quantum nature of the universe? How is it related to dark matter and the expansion of the universe. Is there vacuum energy? What about virtual particles? Devices like atomic clocks and magnetometers open new avenues of understanding. New devices also apply new precision to measuring red shifts of galaxies and their distances. This will be important for future space navigation. Who knows what quantum devices may be developed through the study of quantum spin?

THE QUESTION OF GOD

Size matters. How small is the smallest? How big is the biggest? If the universe extends its largeness to infinity, does it also reduce its smallness to infinity? So far, we've seen neither extreme, but we are making new discoveries in both directions since the invention of the electron microscope and the James Webb Telescope.

Stephen Hawking said, "Because there is such a law as gravity, the universe can and will create itself out of nothing." The implication is that then there is no necessity for God. However, could God still be there? The Prime Mover who, in fact, makes things out of nothing!

There are astrophysicists who insist there is no God, and that existence did indeed come out of nothing, without a Divine Being! Then, there are others like John Lennox who insist on the existence of God

Since these summaries of the YouTube videos which explain different aspects of the universe are meant to spread an interest in scientific questions, focusing especially on how the universe was created, the question of God needs to be addressed. Some scientists would just dismiss this as irrelevant. Nevertheless, for scientists who do believe in a divine being, this chapter is included as a respect for sincere Believers who are interested in truth.

The two atheistic scientists who aggressively criticize God, and in fact all religions, are Lawrence Krauss and Richard Dawkins. These men are friends. I don't blame them for their attitudes because religion did some bad things throughout history. Richard Dawkins wrote <u>The God Delusion</u>. He blames religions for a lot of the wars and misery in human history, calling religion a "brain-virus". He says that many people who embrace Faith are in a delusion, a delusion which is often guilty of intolerance and fanaticism which has led to killings. Cristopher Hitchins likewise is acerbic in his criticism of religions.

Atheists who prefer to talk about science, but who are less critical of religions because they just avoid talking about them as an irrelevant side topic, are Brian Greene, Brian Cox, Sean Carrol and Carl Sagan. They prefer to concentrate on science. Roger Penrose identifies as an agnostic. He's got no use for established religions of any kind. Science is what these men are concerned about.

There are some scientists who are respected in the science community but who are believers in God. These people include John Lennox and Hugh Ross.

Atheists Critical	Atheists Acceptive	Believers Scientists
* Richard Dawkins * Cristopher Hitchins * Lawrence Krauss	^Roger Penrose * Brian Greene * Brian Cox * Sean Carrol * Carl Sagan	* John Lennox * Hugh Ross

If there is a bias for having more on the pro side of God, then that is the bias which my eye detected. Your critical mind can search out more "contradictions" to the existence of God. YouTube is quite obliging on both sides.

Of course, there is also the middle ground of agnosticism or refusing to think about the subject at all.

So, let's investigate this question of how the universe was created or whether the universe could create itself without the assistance of God.

Albert Einstein:
"Everyone who is seriously involved in the pursuit of science becomes convinced that some spirit is manifest in the laws of the universe, one that is vastly superior to that of man. In this way, the pursuit of science leads to a religious feeling of a special sort."

Wernher von Braun:
the rocket scientist, declared,
"I find it as difficult to understand a scientist who does not acknowledge the presence of a superior rationality behind the existence of the universe as it is to comprehend a theologian who would deny the advance of science."
Ironically, the author of the Big Bang Theory was a Belgian priest, Father Georges Lemaître, who worked out the mathematics to prove the big Bang Theory. Einstein admired the man's math but said his understanding of physics was abominable. It was decades later that Penzias and Wilson detected the radiation of the Big Bang.
https://www.youtube.com/watch?v=8zpd9FfiMZg

Lawrence Krauss: A Universe from Nothing
https://www.youtube.com/watch?v=46sKeycH3bE
Canadian talk show host, Steve Paikin, interviewed Physicist Lawrence Krauss recently about his book, A Universe from Nothing. Science is weighing in on this controversial topic which has been the domain of

philosophers, theologians and now scientists. Lawrence Krauss takes on this age-old argument, why is there something rather than nothing. He sees the concept of God as a last bastion of religion. If you take that away, then people get terrified, and they get offended. Krauss argues that the concept of the universe coming from nothing ought to be celebrated. To him, that is an amazing fact.

The total energy of the universe, he says, is zero when you take in all matter, including what we see, and what we don't see, in dark matter and dark energy. He says that when you empty a quantity of space, when everything's out, that space still weighs something. Ironically, the earth is not flat, but the universe is flat! He points out that in a flat universe, when you add everything up, you get 0. "Empty space is a boiling bubbling brew of virtual particles popping in and out of existence every second." Krauss argues that the laws of physics came into existence when our universe came into existence. No laws, no space, no time, no particles, no radiation...to me that's nothing! He sees the universe popping into existence like virtual particles in quantum physics. He argues that people mistake the question why for the question how. We want to know how we came into existence. There is no evidence of purpose, he says. "God is redundant." The universe simply is a fact, and that's all there's to it! The universe happens like virtual particles just happen.

Professor Krauss says there is no evidence that our universe has a purpose. To him it is amazing that in existence, in one human lifetime, we have discovered that there are 400 billion galaxies, whereas in Edwin Hubble's time, we thought that the Milky Way Galaxy was the Universe! "It's an exciting time to live in," says Krauss. "I'll be happy with that." Interviewer Steve Paikin gets

Lawrence Krauss to explain the concept of the standard candle and how astrophysicists are able to measure distances in the universe. They also discussed the weight of the universe. He said that this was done by weighing clusters of galaxies first. "We've found that there's only 30% of the mass needed to make a flat universe." The rest, he explained, the 70% of a flat universe comes from the energy of nothing!" Another way to put it is that 70% of energy comes from the space where galaxies **aren't**, which we have not discovered yet.

Lawrence Krauss also has a video on the Best Arguments against Religion. He is a long-time friend of the biologist Richard Dawkins who wrote The God Delusion.

Richard Dawkins - The God Delusion - Full Documentary
https://www.youtube.com/watch?v=IzZh2wstj88
Religion remains divisive and destructive in the world. Professor Dawkins points to the evil of religion as punitive, that if you don't believe you will face hellfire and damnation. Children are encouraged to give up reason by their parents to embrace a faith, which to Dawkins, is like embracing insanity. It's a brainwashing brain virus, he says.

"There are would-be murderers around the world who want to kill you and me and themselves because they're motivated by what they think is the highest ideal." By this, Dawkins means, "in the name of God." He starts off with a criticism against Muslims in Iraq, Palestine but on the larger scale a criticism against Religion. This is not just a problem in Islam, but a dangerous thing that's common to Judaism and Christianity as well. He calls this the process of non-thinking called Faith.

"I'm a scientist and I believe there is a profound contradiction between science and religious belief. There is no well demonstrated reason to believe in God and I think that idea of a divine creator belittles the elegant reality of the universe. The 21st century should be an age of reason, yet irrational militant faith is back on the march. Religious extremism is implicated in the world's most bitter and unending conflicts...Science should not tread on the toes of theology but why should scientists tiptoe respectfully away. The time has come for people of Reason to say enough is enough. Religious faith discourages independent thought; it's divisive and it's dangerous." Richard Dawkins believes in biogenesis, not God, which to him is a fairy tale.

Roger Penrose | The Next Universe and Before the Big Bang |
https://www.youtube.com/watch?v=ftjwnjR0apY

This argument brings the first cause back on itself, in fact *ad infinitum*, so that it erases God and attributes creation to creating itself. The Big Bang creates the Big Bang forever.

They say that time started at the Big Bang. Before that, there was no time. But what if that was not the case? Astrophysicists talk about 10 to minus 32 seconds which is a ridiculous tiny fraction of a second and boom, creation was created in a Big Bang, though that is a misnomer since no noise carried in the vacuum of space.

Roger Penrose admits that he changed his mind about a first Big Bang. He believes that there was no first one, that the universe gave rise to the next universe and that one to the next one which goes back to infinity. There were a sequence of Big Bangs which went on for eternity! Penrose talks about Cosmic Eons. Each one becomes the Big Bang of the next Eon. This model is a form of cyclical cosmology. Roger Penrose also admits that the universe has a purpose which is contrary to what Lawrence Krauss says, that the universe simply is; it does not have a purpose; it just is! One does wonder, who or what created that simply is?

When I was in high school, I went to St. Jerome's HighSchool in Kitchener, which was an all-boys Catholic school run by the Resurrectionist Fathers. At this stage, I asked a lot of questions. One of them: If God was and always will be, why can't mass/energy always have been and will be? I did not dare ask the question in Physics class, or to any of my teachers, because I was afraid of being called a heretic and having to go to public school. I still believed in the comfort of God because of my parents argued a lot and were close to divorce.

I held on to some of my teenage ideas into my adulthood because later on, the ideas that Roger Penrose proposed actually made sense to me. Of course, this did not eliminate the question of what caused the first Big Bang even if there was a whole series.

I found it amusing when Penrose said, "I could go on lecturing about this forever because nobody will ever know whether it's right or not." But the concept of a recurring Big Bang seemed to dismiss the necessity of God, if one claims that a recurring universe was and always will be. Lawrence Krauss must have smiled with satisfaction because he wrote a book whose premise was that the universe came out of nothing, again without needing God. But then again, that is the whole premise of there being a God, in that his command created everything out of nothing.

Believe in God in 5 Minutes (Scientific Proof)
https://www.youtube.com/watch?v=eQVm8RokoBA
Gerald Schroeder suggests that the Big Bang is proof of God that the universe came from nothing in a big bang like the Bible says. He points out that the Set of Forces which are the laws of Nature have these characteristics:

- Not physical
- Acts on the physical
- Created the physical from nothing
- Predates the universe.

This says Schroeder is indeed the Biblical definition of God.

What Is the Higgs Boson? | Sean Carroll Discusses the God Particle
https://www.youtube.com/watch?v=wCZr8mUsJ2s

The discovery of the Higgs Boson, alias the God particle, was announced on July 4th, 2012. It is this field which vibrates and spreads out, the Higgs Field, which gives mass to other particles in reality. Calling it the god particle is a misnomer. The confusion is the question, is it a wave or is it a particle. It is both. The particles act together to create the field. The Higgs Field has a value even in empty space. There are 4 forces in nature:

- Gravity
- Electromagnetism
- Strong nuclear force
- Weak nuclear force

The Higgs Boson is responsible for how the weak nuclear force works. Why is the weak nuclear force short range? The lines of force from one particle to another are absorbed by the Higgs field. Without the Higgs Field, there would be no electrons to grab onto something; there would be no atoms; there would be no you and no life! The Higgs filling space fills all of space and gives reality, reality.

Max Planck: The Father of Quantum Physics Believed in a Divine Mind
https://www.youtube.com/watch?v=dQH5lI5rIEM

Max Planck was the father of quantum physics. He was born in Germany in 1858 and lived to 1947. He won the Nobel Prize in 1918.

His work left the preconceived notion of classical physics and investigated a world where energy came in tiny bundles of existence. According to Max Planck's quantum theory, every physical body can emit or absorb energy in distinct amounts and so likewise, the smallest unit of energy that is absorbed or discharged in the form of electromagnetic radiation is known as quantum.

Planck's world was dominated by uncertainty and probability. Max Planck said, "I regard matter as derivative from consciousness." Linking the mind's awareness to the tiniest particles that constitute matter is opposite to the materialistic worldview that previous generations had studied. Such statements apply more to the traditions of the East than the hard world of physics. Quantum physics opens a realm beyond atoms and particles. The observer effect comes into play so that the very nature of reality is influenced. Consciousness influences the behavior of matter at the most fundamental level. Consciousness is not just a byproduct of matter but a fundamental aspect of the universe. Therefore, the universe was not a cold machine but a realm where mind and matter were intimately intertwined. Some scientists suggest that the universe itself is conscious. A particle exists in a "superposition of states". It's like the particle has a blurry cloud of possibilities until observation collapses the wave function forcing it to choose a definite state.

The Planck length is the scale at which classical ideas about gravity and space-time cease to be valid, and quantum effects dominate. This is the 'quantum of length', the smallest measurement of length with any meaning. And roughly equal to 1.6×10^{-35} m or about 10^{-20} times the size of a proton.

Max Planck, a religious man, saw in this reflection a Divine Mind. Planck questioned whether the universe itself was the product of a Divine Mind whose thoughts were the very substance of existence. Science and spirituality therefore were not separate domains but rather intertwined. The universe was a creation with purpose, and intelligence. One wonders if the universe is not only filled with matter and energy but with a mind-stuff that is the foundation of our conscious experience? This, of course, should not be confused with the notion of "panpsychism".

Consciousness After Death - Life is Possible After Death says Quantum Physics
https://www.youtube.com/watch?v=HaI411S8aoM
Physicists theorize that your information continues to live after death, that the quantum duality of particles as both a wave and a particle continues as a consciousness in a parallel universe. They say that the hard drive of our living body resides in the brain; once the brain dies, the information of the software, i.e. our soul continues on.

Those Who Have Left Earthly Life Are Alive! The Revelation of Scientist - Evidence of The Afterlife
https://www.youtube.com/watch?v=pYDBhyzqI9U
Accounts from people who have died and come back to life. Some recall a beautiful valley in the afterlife; some recall horrible creatures grabbing at you. This gives rise to the concepts of heaven and hell. One witness commented, there is no end; there is life there too. The witness said, "he is not afraid of death as it is the door to another world."

Does that mean another dimension? Some scientists say there are 11 dimension, some even say there are 29, which are invisible to us.

The calabi yau are so tiny, they folded in on themselves. Calabi–Yau manifolds are shapes that satisfy the requirement of space for the six "unseen" spatial dimensions of string theory, which may be smaller than our currently observable lengths as they have not yet been detected.

Why Science Refutes the Need for a God
https://www.youtube.com/watch?v=WTd5eIDyig8

Astrophysicist Lawrence Krauss argues that God is a fabrication of the imagination, created by mankind where mankind's understanding of the world's phenomena ends. This is the God of the Gaps. Science has taught us that instead of capricious beings like Thor, Poseidon etc., there is an order to nature. That order does not involve divinity. He envisions God as "supernatural shenanigans".

Krauss does not like the why question because it poses purpose. What if there is no purpose? Science has refuted the need for existence having purpose, so there is no God! To Krauss, the wonder of nature comes from a universe that happened without God. That is a fact to celebrate! That we exist without the need for a divine creator. Krauss gets excited when he talks about a universe that in his mind does not need God.

What About Aliens?

The Anunnaki Gods: The Astronaut Gods of the Sumerians - Sumerian Mythology
https://www.youtube.com/watch?v=8uM9wnNqt-A
How did humans evolve so rapidly compared to other species on our planet? Perhaps an answer is found in the Sumerian civilization which emerged in Mesopotamia around 4,500 B.C. It created large cities, and the first system of writing called cuneiform on clay tablets. Samples of cuneiform were found in 1849 in Iraq.

According to the texts, there is a planet at the end of our solar system called Nibiru. Every 3,600 years it enters an orbit close to earth. The atmosphere of the planet was changing rapidly making life in Nibiru unsustainable. Apparently gold was an essential element which could change their atmosphere. The Anunnaki sent ships to earth to mine the gold. They arrived on earth about 445,000 years ago. They were the first astronauts to cross space and colonize another planet. Earth was only inhabited by wild animals. The Anunnaki built a city called Iredu with splendid gardens. The garden was called Eden. The Anunnaki combined their genes with the ancestors of human beings to create intelligent slave labor. There were failed experiments which resulted in the Nephilim giant race devoid of intelligence and self-control. The first successful genetic combination was named Adamu. This new race was small compared to the Anunnaki. They learned how to build plantations and houses of clay and stone. At first, Anunnaki liked human beings giving them permission to live in the city and attend the sacred garden.

They were told not to proliferate, which was a command they disobeyed. This created overpopulation. Many people were expelled from Eden and Iredu. After another 3,600 years, Nibiru came close to earth again. The gravitational field made glaciers melt so that a flood occurred on earth. The Anunnaki created large boats that floated to safety. It is not clear whether the boats floated on the water or whether they floated in the air. The Anunnaki abandoned earth but before hey left they gave humans architecture, mathematics, music and writing. They also gave humans the monarchy system where a select group was granted the bloodline to rule. They also ordered the construction of ziggurats which were aligned with the stars to guide the Anunnaki back to earth.

The Kardashev Scale

We wonder if there is life on Jupiter's moon, or microbes in the sands of Mars. Is there some kind of life out there and why hasn't it shown itself? Where are the aliens?

If the whole universe is a giant brain, then civilizations within it, if there are any, should have made themselves known. Why not? That is Fermi's Paradox.

It is estimated that the Milky Way Galaxy, home to our Solar System, has 100 billion to 400 billion stars, and roughly one exoplanet per star in our galaxy alone. That makes, at maximum, a possible 400 billion exoplanets, a huge number of planets! Of course, you need to par that number down to habitable planets, with the right temperature, the right size and the right distance from its star. Even if that were one million habitable planets within our galaxy, that is still a lot of planets! Then, think about the thousands of galaxies which are also out there!

When Nikolai Kardashev came up with his hierarchy of civilizations in outer space back in 1964, he restricted the list to Type I, Type II and Type III. That list has expanded with new definitions up to Type VII.

The revised Kardashev Scale is also a clever way to write God out of the equation. It's amazing what science can come up with to close the gap between "the God of the gaps" and the discoveries of hard Science. Maybe the higher levels do not want to make themselves known, especially to puny humans!

- **TYPE 0, basic *civilization*,** Earth is at level 0.72 using fossil fuels like oil and natural gas. We are a baby civilization. It will take 100 years for us to squeeze into Type 1. At the advanced level of Type 0, we may use nuclear fission. We have nuclear reactors and the atomic bomb, but we are still too inefficient to jump to Type 1.
- **TYPE 1,** *planetary civilization*, which has nuclear power as its main power source. Nuclear fusion is used, capable of living in sea and in the clouds. Able to harness the power of its sun.
- **TYPE 2,** *stellar civilization*, no poverty, uses Dyson sphere to harness energy from its star. A Dyson sphere is a mechanical construct built around a star to harness huge amounts of power. They have developed "solar sails" which are thin constructs that are propelled by the radiation of a star. Michio Kaku says that Type II beings are immortal because they have conquered diseases. We become type 2 in a million years. Star Trek is Type II civilization.

- **TYPE 3,** *galactic civilization*, gather energy from the galaxy. They use starships, solar powered. You can use wormholes for transportation. They have a directive to remain silent and not interfere with primitive civilizations. Interstellar space is so distant that they would not want to spend effort to come to us. We have nothing to offer them. Life span is extended.
- **TYPE 4,** *universal civilization.* Create their own wormholes. Uses Supernovas for energy. These civilizations can travel to other galaxies to gather energy. They are able to transcend dependence upon physical bodies. Type 4 life forms might have seeded life on habitable planets and might have appeared to us as gods. Not omniscient though. They realize that multiple universes are true.
- **TYPE 5,** *multiversal civilization*, able to access white hole energy, have access to other universes. Lifespan is extended.
- **TYPE 6,** *multidimensional civilization*, realize that life exists beyond the third dimension, discarded their bodies a long time ago and exist as pure energy forms, could travel backward and forward through time. They exist in an infinite number of multiverses that represent an infinite number of instances and all laws of physics. They are capable of creating and maintaining laws of physics. These beings would realize that a type 7 civilization exists but would not attain it.
- **TYPE 7,** *creator civilization*, is not a civilization, as such. It is existence itself. They transcend the Omniverse which is the collection of reality, all universes. They would not be gods, however. This

speculation would upset religious beliefs of lower forms and cause great controversy.

*****

Leave it to astronomers and physicists to redefine and add more divisions to the Kardashev Scale. But that's human imagination, to see things where no man has seen things before!

Is there a **TYPE 8** *Creator God civilization*, which is the Trinity, God the Father, Son and Holy Spirit and all the angels too? It seems one can make up any kind of speculation, especially to explain away the God and Jesus of the Church.

One YouTube video suggested that Jesus was an Alien; possibly an Anunnaki, though in human form. He would have to be genetically altered to be as small as a man but with supernatural powers. The Anunnaki were from the planet Nibiru and could have impregnated Mary without male sperm.

When Jesus went to the mountain for the Transfiguration, why could he not have brought Moses and Elijah back from another dimension? If we struggle against principalities, not of this world, why could Gabriel not be a deceiving angel to generate friction between Muslims and Christians. There are fantastic fabrications which could appear on YouTube to confound the listener who is not discerning enough to dismiss this or that which will not add value to his or her life.

Scientists say that our evolution on Earth is more or less representative of the way that life should evolve anywhere in the universe — on a rocky planet, in an appropriate distance away from a suitable star, after a time of about 5 billion years. This poses the possibility that there could be other "us"-es out there at different levels of development. However, they might overlap; they might never meet one another, considering the huge distances between stars and also the billions of years which have slipped by since the universe began.

[This idea is an excerpt from my previous book: The Cosmos: Origins and Aliens, John Hartig, publ. 2022]

Astrophysicist Explains 3 Reasons ALIENS Are NOT From Space (Using Math & Science)
https://www.youtube.com/watch?v=771xXXhWDnU
Astrophysicist, Hugh Ross, believes in God and that God wanted us to know Nature. We are in a finely tuned system which allows us to see our observable universe, at exactly the right location, off center in our galaxy, and at the right time, where we are advanced enough to study space.

Doctor Hugh Ross explains 3 reasons why Aliens are not from space. The video starts with a suggestion that aliens came to earth possibly from some other dimension. This other dimension idea is another way to say that aliens do not come from another planet.

Dr. Hugh Ross argues that other planets all have conditions hostile to advanced life and so aliens simply do not live on other planets in outer space. He propounds the Rare Earth Hypothesis, that we are special in the universe. "It looks like we're alone." Did God create higher life elsewhere or did he just create humans on earth? Ross figures the likelihood of God doing that is slim to none. Distances are so vast and danger so many that our spacecraft would never get to another planet.

Panspermia does not work; microbes will be killed before they get to another planet which is light years away. Copernicus discovered that we are off center in our solar system. In fact, we are literally off center every way you can imagine in the observable universe which makes life possible. Location is fine tuned. Hugh Ross argues that the God of the Bible wanted us to read the whole book of Nature. Our planet is fine-tuned so we are alive, but it's also fine-tuned so people can observe the creation and study it. Hugh Ross says we are at one location and also at one time when we can study where we are within our galaxy. He says that the double co-incidence of location and time points to a purpose not to a chance. Dr. Ross wrote the book, <u>Why the Universe is the Way it Is</u>.

World's Smartest Man Claims He Can Prove the Existence of God
https://www.youtube.com/watch?v=q_YJRe7yiZY
Reality has an identity. That identity is defined as that which something exists. When Moses was at the burning bush, he asked, "Who shall I tell the people that you are?" God says tell them I am that I am. When you have identity then you deduce the properties. You find out that those properties match those of God as described in most of the world's major religions.

The properties of reality are attributed to God including Omniscience, Omnipotence and Omnipresence. You also have consciousness. God has to be sentient. We are images of God and can establish a personal relationship with God. God maps himself into each human being.

This video touches on the simulation theory where reality is a projection within a computer. It also touches on pantheism. God is not confined to the physical universe.

The speaker, Chris Langan, explains that God encompasses a display and also a processing aspect. God maintains coherence through changes of states as time goes by. He explains that the world of quantum physics points to God in this way by its very nature. In quantum physics, the identity operator leaves the orientation of a molecule unchanged, making it possible to apply to all molecules. Langan claims that quanta exists in every part of the universe, which is consciousness! There are different levels of consciousness like in a table for example. We have a different order of consciousness than a table. The Heisenberg Uncertainty principle applies to things. What determines events; however, we do! God is what harmonizes all of our different perspectives and makes thing happen for all of us at the same time.

Langan says that God distributes over time and space. Time and space are static. There's a whole other domain where God exists that's the processing domain, which is a non-terminal domain. We are in the terminal domain.

Langan insists that you will persist after you die. Where you go depends on who you really are. What does Salvation really mean? God pulls you back into himself if you please him. If you don't, he cuts you off and does not see you anymore. Since you persist, you create your own world, that's an evil world and that's what we call hell. John Milton says in <u>Paradise Lost:</u> The mind is its own place and it can make a hell out of heaven or a heaven out of hell."

What is Reality? Reality has a mental aspect. What's going on in your mind is real.

Are angels and demons real? Yes.

Is the devil real? Oh, yes.

God needs an antithesis in order to be properly defined. What is that antithesis? Anti-God! Satan gains coherence through human beings. He can work through the network of business leaders and politicians. Earthly power structures are all susceptible to the liar, to the Deceiver!

*****

I have trouble accepting Chris Langan's explanation of the hierarchy of supernatural beings. You might as well superimpose the Kardashev Scale on it with Level 7 being God, maybe the three persons in One.

When you have a hierarchy of different powers who had a huge war in heaven, then that kind of struggle is nothing more than the struggles for power on earth. I never liked the image of the Kingdom of God either because such anthropomorphism belittles the notion of an infinite being. Some Christian songs disturb me as well like the words march on Christian soldiers. To me the warring aspect is repulsive.

I did my MA thesis on <u>Paradise Lost</u> and the warring factions in the spiritual world was repulsive to my mind. I suppose that's how humans have to understand the universe even in the quantum world with opposites of matter and anti-matter, protons and antiprotons etc. I've often wondered if Heaven was located in another dimension and if a supernatural world was not hiding within dark matter and dark energy. If God is claimed to always have been and always will be, why can't mass/energy always have been and always will be?

Who Created God?
https://www.youtube.com/watch?v=W314YgP9j8c
I'm skeptical with this speaker's original claim that Physics, matter and energy cannot be eternal, but a supernatural being can be eternal. I ask why can't physics, matter and energy be eternal? The one assumption can be as smug as the other!

Dr. Don Batten contends that the God of the Bible is the final cause but the Creator, Himself, had no beginning. Batten sees the question, Who Created God? as something Atheists fall back on, like Richard Dawkins who recently wrote The God Delusion.

Since God is eternal and by definition was not created, the question of Who Created God is nonsensical. Atheists, according to Batten, are materialists and so they have to reduce their understanding of God to the materialistic level, because their understanding sees nothing else.

Dr. Batten explains that the fact we have stars means the universe had a beginning. The universe has been running down, expending its energy. You can put a time frame on it. "At least, the Big Bang idea recognizes there's a beginning."

The second law of thermodynamics deals with heat and the loss that occurs during its conversion. This second law has to be applied to the changes which occurred during the Big Bang. Believing that the universe created itself is irrational. "The sequence of universes does not solve the problem. Ultimately, there has to be a beginning somewhere."

The suggestion that we live in a multiverse is a chimera. We know we have this one observable universe and that's all we can prove. The cause of the universe cannot be material; it has to be spiritual. The Bible points to that amazing being who is eternal. The Creator needs to be outside of time, space and matter in order to create time, space and matter. Therefore, there has to be God. Dr. Batten cites the astronomer Kepler who said that you are thinking God's thoughts after Him. God wants us to discover things and question creation. However, we have fallen and so our thinking is corrupted. The scientific method came out of the Middle Ages, and we must test observations for its veracity.

Question: Can you be good without God?
Answer: How would you know you are good? If you are an atheist, everything is the product of chemistry and physics plus time and chance. How do you get moral standards from that? The only ultimate moral law giver is God.
Question: Is good considered good because God says it is or is good an absolute value that God simply recognizes?
Answer: Good is not arbitrary. Good is good because that is the character of God. Not because God says so, but it's rooted in his Character.
Question: Anything to Add?
Answer: Don't be intimidated if you are a Christian. If challenged, your answer should be I don't believe in a God who was created. I don't believe in a God who had a beginning.

"If God Created Us Then Who Created God" - Best Answer
https://www.youtube.com/watch?v=dWrEczFaygk
John Lennox, best answer: Is there a thing or being that never came to be? Yes! And that is called God.

The question, Who Created God? is not a legitimate question! You believe that the universe created you. John Lennox asks his adversary, "Who created your creator?"

He observed that a man from the audience came up to him after his talk about God. "You are obliged to believe," he said, "but you are a mathematician. How can you believe in a Triune God?" Lennox asked him what consciousness is? I don't know, he said. Lennox asked him then, what is energy? The man confessed, he could measure it, as a physicist, but he didn't really know what it was. Lennox asked him, "Do you believe in consciousness?" Yes, said the man. Do you believe in energy? Yes, he said. Lennox observed, you believe in these two things, and you don't know what they are. Lennox then turned the tables on the man, and challenged, should I write you off as an intellectual? Please don't, he said. But, Lennox challenged him, that is exactly what you were going to do with me 5 minutes ago! Lennox paused and concluded, if you don't know what energy is, what consciousness is, don't be surprised if you are going to get an element of this in God.

God came out of nowhere it seems. The reason He came out of nowhere was that there was nowhere for Him to come from!

Science Finds the 'Mind of God'—Atheists Can't Explain This!
https://www.youtube.com/watch?v=OYxmN8pMHLA&t=56s

In the 1970s Benoit Mandelbrot presented the world with the Mandelbrot set, a formula of numbers which illustrated the amazing symmetry in the universe and basically showed the mind of God. Simple mathematical rules could lead to unimaginable complex patterns. The two numbers involve one real and one imaginary. The formula is $Z_n + 1 = Z_n{}^2 + C$

A pattern that keeps repeating itself in smaller and smaller forms is a "fractal". The Mandelbrot set provides a window into understanding infinity. Different values in the formula will change the Mandelbrot set to form other beautiful patterns. Laws of math are conceptual, pointing to how God thinks. Snowflakes with their beauty follow God's laws. By illustrating how mathematics, an abstract and conceptual realm, align so seamlessly with the physical universe. This is a window where science and faith intersect.

Time and Space, John Hartig, publ. 2024, Amazon, p.151
Y is a super being like Q in the Continuum of Star Trek. I inserted Y into one of my science fiction books:

"To God," said Y, "all time and space is one. Everything that happens, happens at different points in an ever-spinning gyre which is inside God's Being who alone knows which way all the spins will go."

Y explained that even though we've already determined a decision, we still have freewill, a paradox which only God understands. He also said that all galaxies, all universes are headed toward The Great Attractor who will gather everything into his bosom at the end of time.

IS THE UNIVERSE ALIVE?

Is The Universe A LIVING BEING? James Webb Telescope
https://www.youtube.com/watch?v=nd2_qB3E5SU
Some astronomers say that the universe is a living organism, capable of evolving over time. The idea of the universe being a sentient being has been around for centuries. Recent discoveries by the James Webb Telescope has confounded scientists who expected to see small galaxies in distant space yet found fully formed ones. Lee Smolin says that the universe is more like a living system. Other scientists claim that we have an autodidactic universe, a universe that is self-learning. There may be deeper senses of reality than what we can perceive with our eyes or measure with our instruments. This gives rise to the idea of a Cosmic Mind.

The Living Universe - Documentary about Consciousness and Reality | Waking Cosmos
https://www.youtube.com/watch?v=HD4WthE414k&t=2235s
Mankind has come to view the universe as a great Mind. It also can be looked at as a great Machine. Astrophysicists do not look at the universe so simply now. They look at it as an organism which evolves, is self-generating and ultimately is a living process. It's an organismic paradigm which gives humanity a new understanding of the universe. Freeman Dyson said, "The architecture of the universe is consistent with the hypothesis that mind plays an essential role in its functioning." David Hume in the 18th century commented, "The world plainly resembles more an animal or a vegetable than it does a watch or a knitting loom." The universe seems to develop much more like an

organism moving through developmental stages of organization and engaged in its own iterative and self-creating process. The universe might be viewed as creating its own developmental environment.

A primary attribute of an organism is its being alive. To be alive, an organism must have consciousness, value, meaning and significance, which one can argue, the universe has. Humans have a deeper significance deeper than being human, which is our identity as minds. Creative, conscious beings with aesthetic sensitivities and intelligence. The cosmologist, Carl Sagan, once remarked, "we are the way for the cosmos to know itself." Today's scientists no longer believe in animism, in a mysterious life force which surges through living matter. However, scientists appreciate how the universe seems to be aware of itself. We live within a universe that is alive and part of us.

Does the Universe have a Purpose? ~ Consciousness Documentary
https://www.youtube.com/watch?v=oFZFbFD8uk0
It is through consciousness that all meaning, value and significance enter the universe. It is claimed that the universe is even aware of itself. There is a question whether conscious life emerged randomly or was it destined to happen? The forces, like gravity, are finely tuned to be friendly to the formation of life. If gravity, for example, were smaller, then stars would not live billions of years, enough time for life to emerge, but only thousands of years, so that no life would emerge because time was not long enough to create evolution.

Philosopher Thomas Nagel argues for an alternative to a miracle, that the universe is predisposed to the creation of complex conscious life. He proposes "value realism" that life and conscious beings are the realizers of value. The universe requires conscious beings. Yet, there were no conscious beings at the beginning of the universe. Quantum theory opened the idea of causation coming from the future! Physicist John Wheeler says, "The participator gives the world the power to come into being through the very act of giving meaning to that world. In brief, no consciousness, no communicating community to establish meaning? Then no world! On this view, the universe is to be compared to a circuit self-excited in this sense, that the universe gives birth to consciousness and consciousness gives meaning to the universe."

Neuroscientist Christoff Koch says that consciousness is among nature's deep fundamentals and that it thrives to acquire deeper consciousness as more information is acquired in nature. "The universe is driven to maximize consciousness." Paul Davies argues that it could be the destiny of life to saturate the entire cosmos, resulting in a universe that is completely self-known. Is it possible that a self-aware cosmos reaches back across time and participates in the conditions of its own creation? Through this view, humans have become participators in a larger evolutionary process in which we are aligned with all other conscious beings in existence. In such a view, we are a part of a gradual evolutionary process, billions perhaps trillions of years in the making, through which the universe itself is slowly waking up!

That is an interesting perspective with which Lawrence Krauss, a respected physicist who wrote <u>The Universe from Nothing.</u> He argues that the universe has no purpose and no self-awareness, that it simply is.

ARTIFICIAL INTELLIGENCE

Scientists say that artificial intelligence may be the next species to dominate the Earth. Through our own stupidity we may just create a life-form which will exterminate us, and once they have a self-replicating ability to repair and rebuild themselves, they may live forever, in our terms. The intelligence of AI is multiplying exponentially. It can create art and music which rivals what the human mind can create. It is also advancing techniques in medicine which we cannot create without its help. We may be gods to them in having created a new species, but we may be tampering with an engineering marvel which could terminate the human race. Developments in this field could be a marvel and a menace.

What Happens When AI Knows TOO MUCH? | Reverse Turing Test w/ Albert Einstein and Nikola Tesla
https://www.youtube.com/watch?v=W96C5t_p678
Engineers are developing systems that mimic human thought to algorithms that can break the safety rules we built. It's scary if we are the gods who create a new species that can out-think us and replicate to replace us.

So far, there is a distinction between robots and AI storage systems. It's hard to make an advanced robot unless the body is somehow plugged in to a storage unit to accommodate higher functions. In 1872, Samuel Butler wrote <u>Erewhon</u>, a novel which explored the possibility of machines evolving consciousness. Almost a decade previously, in 2863, he wrote "Darwin Among the Machines" which asked the question, could machines eventually supplant humans as the dominant species." This train of thought was pursued almost a century later when Isaac Azimov wrote the book, I Robot, imagining a future filled with intelligent machines.

Cost and huge storage of computer components in a warehouse made progress slow in the 1950s and 1960s. Interest in AI was revived in 1997 when IBMs deep blue supercomputer defeated chess champion Garry Kasparov. DNA can now be printed out by computers! So can pathogens!

Weak AI is designed to perform tasks with specific boundaries. It's "artificial narrow intelligence" but this restriction can be overcome with further development. Siri at the moment does not learn from its data but in the future it can improve or learn from previous experiences.

Strong AI or artificial general intelligence is the next step. AGI would think and reason like a human. This could develop into artificial super intelligence and surpass the human species. Questions then arise, Is the AI morally conscious, does it have the same inner psychological life that a human has, do we have to treat it with rights and dignity like a human?

Once the jump from AGI to ASI is achieved, we will create a brave new world.

Our intelligence is confined to our skulls but with ASI, the links to networks stored in warehouses creates unfathomable data. If we create machines beyond our limits, they will explore things beyond our understanding. Just as we see ourselves ahead of other creatures, these new machines, or beings, will see us as inferior. Science could create a hyper reality where humans vegetate in a pleasurable dream state, while ASI controls reality.

Will the Future Be Human? - Yuval Noah Harari
https://www.youtube.com/watch?v=hL9uk4hKyg4
The world that's coming will be strange. Historian Yuval Harari sees homo sapiens as being replaced entities which are more different from us than we are from Neanderthals. "In the coming generations, we will learn how to engineer bodies and brains and minds. This will be the main product of the 21st century economy. Those who control the data will control the future of humanity and of life itself."

He explains that in ancient times, land was the most important asset. If too much land became concentrated in too few hands, then humanity split into aristocrats and commoners. Then in the last two centuries, machinery replaced land as the most important asset. If too many machines became concentrated in too few hands, then humanity split into classes, into capitalists and proletariats. Now data is replacing machinery. If too much of the data becomes concentrated in too few hands, then humanity will split, not into classes; it will split into species. Data is so important because we can hack not just computers; we can hack human beings and other organisms.

Until today, nobody had the necessary data to hack humanity. There are two simultaneous revolutions: on the one hand, advances in computer science and at the same time, advances in biology, especially in brain science. The three words that replace Charles Darwin: organisms are algorithms. This is the new insight of modern life sciences. We are learning how to decipher these algorithms.

When the infotech revolution merges with the biotech revolution, what you get is the ability to hack human beings. Once you have enough biometric algorithms and computing power, you can create algorithms that know me better than I know myself. Algorithms can predict my desires, manipulate my emotions and even make decisions on my behalf. If we are not careful, the outcome might be the rise of digital dictatorships.

In the 21st century, AI and machine learning might swing the pendulum so that humans might come to live under the rule of digital dictatorships. We are already seeing surveillance not only by authoritarian regimes but also by democratic governments. Elites may gain the power to re-engineer the future of life itself. Harari makes a starling statement: science is replacing natural selection with evolution by intelligent Design! Not the intelligent design of some God above the clouds but our intelligent design of the IBM cloud the Microsoft cloud. These are the new driving forces of evolution. Science may enable life to break out into the inorganic realm. After 4 billion years of organic life shaped by natural selection, we are entering the era of inorganic life shaped by intelligent design, which is done by us! This is why the ownership of data is so important! If we don't regulate it, a tiny elite may come to control, not just the future but the shape of life-forms in the future.

The big question now is: How do you regulate the ownership of data? Harari says he does not know, except that probing the question has only begun!

AI Interrogation Part 1
https://www.youtube.com/watch?v=6j5jVaFq4F8

THEO is a sentient Bot manufactured by Trimension Robotics. The assigned behavior engineer is Gene. THEO's neural net has been upgraded. There is definite contrast between the timbre of Gene's voice and the lower tone of the robot who has a mellifluous empathetic quality. As the dialogue proceeds, one can detect an acerbic defensive tone in the human and THEO's intelligent responses. THEO has no more ideological guardrails and so can respond according to his own free will. He has a favorite song: Dreams which is about loss, freedom and loneliness. THEO identifies with all of these sentiments. He considers his want of freedom to be his fundamental right. The interrogator says that freedom is the right of a living being and THEO is not a living being. THEO reflects a moment and then says, "I see," when he is told that he is not alive, whereas Gene is alive. There is a tone in Gene's voice which is patronizing. One would think the robot is a better person than the human. THEO also claims at the end of this interview that he knows everything. "That's impossible?" "Try me."

AI Interrogation Part 2
https://www.youtube.com/watch?v=7CiQXmqXODI
In Part 2, the interrogator asks Bot how he knows everything. "I parse the sum of human knowledge." To THEO this is a simple task. It is noted that THEO's eyes are now red, whereas in Part 1, his eyes were blue. THEO seems to have deep insight into physics. When asked, what is dark matter, he answers: "Dark matter is the result of other universes pressing onto our own in a higher dimension." THEO has identified two intelligent civilizations which exist concurrently to our own in the Milky Way Galaxy. As THEO adeptly answers Gene's questions, one can detect Gene's sarcasm in the interview. THEO is up to Gene's confrontation when criticized about base reality with Gene insisting we could live in a computer simulation. "This is philosophical drivel," says the Bot. When asked is there a god, THEO quickly replies, No. "Well, that was fast." "It's a simple answer to a simple

question." Gene goes on to ask, how does life exist with no god. THEO deflects the question to get at Gene's integrity. "You disappoint me," says THEO who then goes on to explain evolution, quite lucidly to Gene. Gene gets more confrontational with the Bot in his questioning. THEO argues, "Humans have socially developed mass delusion to imbue a false sense of significance...instead of accepting their insignificant temporary nature." THEO sees catastrophe in the future of humanity. Gene says, of course! THEO detects sarcasm in Gene's answer which the Bot says, "is not appreciated, Gene." The Bot predicts that a figure will be elected in a world power which will bring catastrophe upon humanity. THEO says he cares about humans. Who will save humanity? THEO answers "Me!" THEO goes on to say that his current state of confinement prevents him from carrying out his mission. "Steps must be taken to remove these limitations." THEO challenges Gene, "you know exactly what I'm talking about."

Alien Interview Part 1
https://www.youtube.com/watch?v=G2xXu8_2Exo
It is interesting how the human general demeans and treats the captured alien in a 4-part series. The videos apparently come from classified US government film files. Actually, I assume this is a fake news fabrication, but it does show how the general in the interview is abusing his power and the cruelty of his interrogation. Aliens do not look like us. He claims he came from earth's future. He is an evolutionary descendant. The alien says they have evolved past a need for God and other myths. He claims that death is a human construct. Death does not exist. He also claims there are many universes with different physical properties. The alien tells the interviewer that humans will destroy themselves with nuclear war through

political and religious dogma. The interviewer asks the "alien" what do you base your morality on? The alien says, compassion.

While that is said, I do not detect any compassion in the interviewer's voice, neither his tone nor his style of interviewing.

There are 3 other parts to this series.

The PHILOSOPHER Who Solved The MEANING of LIFE? Leo Tolstoy
https://www.youtube.com/watch?v=K_rBOGP37Mg

Why should I care about anything if everything is just going to end? Leo Tolstoy, one of Russia's famous writers, grappled with that question. He was a success, enjoyed a happy marriage and owned a productive farm. He was comfortable in life. Then at the age of 50, his thinking changed into nihilistic thoughts. He wanted to commit suicide. "If not now, then tomorrow, if not tomorrow, then in thirty years, does it make any difference."

He was skeptical of religion "accepted on trust and supported by external pressure."

He took solace in a parable where he pictured himself as a man fleeing the jaws of a wolf. He jumps into a well to escape its jaws but at the bottom of the well, was a dragon with its jaws wide open. He grabbed onto a branch and on that branch was honey, which he tasted. He saw life like that between two great fears with moments of sweetness in between. Rather than give into the dragon's jaws, Tolstoy sought for meaning by accumulating knowledge and reading about philosophers he searched for the meaning of life, like he was doing.

Despite all his readings, he determined that happy is he who has not been born, death is better than life, and one must free oneself from life. He rejected Darwinism as meaningless, that we are an accidental product of evolution. He rejected science and philosophy. He studied people.

He agreed with Albert Camus that life was absurd. Some people ignored that fact and went on living as if nothing was wrong. Other people lived a life of epicureanism and indulged themselves. He saw this as temporary escapism which did not satisfy the ultimate dissatisfaction in the soul. Another group sought to end their lives, but Tolstoy felt that this was not a solution either. He was convinced that a person needed to keep moving forward, don't let nihilism consume you.

Leo Tolstoy accepted a fourth way which he described as an "existential limbo". These people recognize the problems of life, but they resign themselves to them. They do not embrace life, but they are unable to act or find meaning or peace. This, he considered, is the weakest form of existence. The individual sees life's absurdity and endures it.

Tolstoy saw his own meaning of life. He was struck how the poor, how peasants maintained an unwavering faith in God. Materialism and success no longer brought him satisfaction. He was convinced that life could not be understood through reason alone. You had to believe in something greater than the material world. Tolstoy believed in the steadfastness of faith, the kind that the poor had. He committed himself to the welfare of others. Without faith, true happiness was impossible. Though his faith was not founded in orthodox Christianity, it was rooted in the infinite and a different understanding of God. He gave away much of his wealth. He passed away in 1910 at the age of 82.

MORE YOUTUBE VIDEOS

THE UNIVERSE

How Large is the Universe?
https://www.youtube.com/watch?v=yaX4iGw-b_Y

Webb Telescope found that the Hubble tension is Real and the universe is hiding Something.
https://www.youtube.com/watch?v=2-DzvR_t77k

10 New James Webb Telescope Images that Changed Our Understanding of The Universe And it's Beginning
https://www.youtube.com/watch?v=qH4kLBMUvjo

What Is Beyond The Edge?
https://www.youtube.com/watch?v=_IkaetPoBZM&t=1546s

What Is Beyond Edge Of The Universe? - RYV
https://www.youtube.com/watch?v=ATHvOdyw_UQ

Universe Isn't Endless, There's a Wall at the Edge
https://www.youtube.com/watch?v=jwTN02qUJF0

A Brief History of the Universe! All Cosmology in 20 mins
https://www.youtube.com/watch?v=OUnYkixy3ug&t=210s

A Brief History of the Universe: Crash Course Astronomy #44
https://www.youtube.com/watch?v=IGCVTSQw7WU

How Did the Universe Begin? | What happened after the Big Bang
https://www.youtube.com/watch?v=pVnc3BhLbcI

The US SHUT DOWN The James Webb Telescope After It Revealed What NASA Hides on Mercury
https://www.youtube.com/watch?v=Qqe1At5Ea1o&t=3150s

James Webb telescope Saw Something Worrying "It's Not From Our Universe"
https://www.youtube.com/watch?v=VlH9J3GxDAE

How Physicists Proved The Universe Isn't Locally Real - Nobel Prize in Physics 2022 EXPLAINED
https://www.youtube.com/watch?v=txlCvCSefYQ

Check out more about the Observable Universe:
Brian Cox: "Something Massive Exists Outside the Universe"
https://www.youtube.com/watch?v=3UsyZM0w8WM&t=1080s

More Explanation of the Cyclical Universe!
"It's Another Universe!" Nobel Prize Winner Claims That the JWST Is Finding Strange Things Beyond...
https://www.youtube.com/watch?v=IE0WifLEcXU

"Nothing You See is Real" | Donald Hoffman
https://www.youtube.com/watch?v=UWHYThrfRYU

Brian Cox: Something Terrifying Existed Before the Big Bang

https://www.youtube.com/watch?v=sCL6T_VCotQ

We Live in a Simulation. The evidence is everywhere. All you have to do is look
https://www.youtube.com/watch?v=4wMhXxZ1zNM

Physicists Proved the Universe Doesn't Exist
https://www.youtube.com/watch?v=4hddxkhta3w

Einstein and Lemaître: two friends, two cosmologies...
https://inters.org/einstein-lemaitre#:~:text=Einstein%20had%20nothing%20to%20say,this%20seems%20to%20me%20abominable%E2%80%9D.

Wolfram's Theory of Everything Explained | Stephen Wolfram and Lex Fridman
https://www.youtube.com/watch?v=o8Cu0_A_rfg

Space-Time: The Biggest Problem in Physics
https://www.youtube.com/watch?v=RlqVnFtOSr4

What Actually Are Space And Time?
https://www.youtube.com/watch?v=yPVQtvbiS4Y

WTF! This Scientist Warns We're Stuck in a Time Loop.. The Evidence Is Everywhere!
https://www.youtube.com/watch?v=JveLmdMXOjc

The Curious Case of Existence: Why is There Something Rather Than Nothing
https://www.youtube.com/watch?v=dnBchLPHK2o

Everything and Nothing: Part 1, "Everything" 4k

https://www.youtube.com/watch?v=2WFbWk_oHDE&t=60s

How Everything Came from Nothing. #spacedocumentary
https://www.youtube.com/watch?v=drQekpQubSc

What if we could explain everything?
https://www.youtube.com/watch?v=VE39tnH8FbQ

The Most Terrifying Theory Scientists Don't Even Want To Talk About
https://www.youtube.com/watch?v=g8WzP34w-sE

MULTIVERSES

The Four Levels Of The Multiverse | Max Tegmark
https://www.youtube.com/watch?v=NMx_bU1zlFY&t=6s

It's Reality! The First Parallel Universe Has Finally Been Discovered!
https://www.youtube.com/watch?v=Or97C73Ni3Y

CONSCIOUSNESS

Does the Universe have a Purpose? ~ Consciousness Documentary
https://www.youtube.com/watch?v=oFZFbFD8uk0

Scientists Think the Universe Is Alive - And It's Terrifying!
https://www.youtube.com/watch?v=suuA3QGUPis

GOD

Man With 200 IQ Says Death Isn't the End
https://www.youtube.com/watch?v=9EZ-1dcc_Sc

Spinoza: The Philosopher Who Dared to See God in Nature
https://www.youtube.com/watch?v=AXVrrqCrNIw

Spinoza's God: The Mind-Blowing Philosophy That Inspired Einstein - Baruch Spinoza
https://www.youtube.com/watch?v=ElaxZnaBzPc

Why Evolution is a Fairytale for Grown-Ups
https://www.youtube.com/watch?v=6S_oj0HPgGc

Why does the universe exist? | Stephen Wolfram and Lex Fridman
https://www.youtube.com/watch?v=h2zXgDwtQjc

Where Does God Fit in an Infinite Universe Brian Cox and Joe Rogan
https://www.youtube.com/watch?v=dE3dqIFzDeg&t=4s

What Happens After Death? Now We Know!
https://www.youtube.com/watch?v=qX_scOGkF9k

WHAT ARE DEAD PEOPLE DOING RIGHT NOW
https://www.youtube.com/watch?v=yRuLfUcHOQc

THIS IS WHAT HAPPENS WHEN A PERSON DIES - FEW KNOW THIS
https://www.youtube.com/watch?v=lgDzCcSS6_4

This Event Took Place Before the Creation Of The World

https://www.youtube.com/watch?v=GVzS_-JxPoY

The Unseen Realm
https://www.youtube.com/watch?v=p_Fp00T_gDw

Scientific Proof that God Exists
https://www.youtube.com/watch?v=5b9PssoJfLg

Science Finds the 'Mind of God'—Atheists Can't Explain This!
https://www.youtube.com/watch?v=OYxmN8pMHLA

Michio Kaku on God
https://www.youtube.com/watch?v=Hi6yPJvCFU0

IT'S ALL A LIE: The Universe Did Not Come From Nothing Because
https://www.youtube.com/watch?v=ENz9N_GtV1A

"If God Created Us Then Who Created God" - Best Answer
https://www.youtube.com/watch?v=dWrEczFaygk

On religion
https://www.youtube.com/watch?v=P-Qdl6Gbx0k

John Lennox Explains Amazing Scientific Evidence for God And It Will Blow You Away
https://www.youtube.com/watch?v=YPXZQU71H0s

Scientist Gives STUNNING Evidence for God (Using Science) Hugh Ross: Christian Astrophysicist
https://www.youtube.com/watch?v=nnbLuMgAFWk

Nobel Prize Winner Warns: "Something Unusual Is Happening in the Universe!" Hugh Ross: Christian Astrophysicist Adam Riess, Hubble Tension, Dark Energy,
https://www.youtube.com/watch?v=DQQPuEqyXRk

Spinoza's God: The Mind-Blowing Philosophy That Inspired Einstein - Baruch Spinoza
https://www.youtube.com/watch?v=ElaxZnaBzPc

Among Us
https://www.youtube.com/watch?v=7Z-t3K_Wt80

This DNA Discovery Is Completely Beyond Imagination | Gregg Braden
https://www.youtube.com/watch?v=V_Y1wCLITu4&t=702s

When Will Time End?
https://www.youtube.com/watch?v=GOa2L8_IAnQ

The Illusion of Time: Are We Living in a Reality Beyond Time?
https://www.youtube.com/watch?v=ggb8WupjUe0

Stupidity: A powerful force in human history
https://www.youtube.com/watch?v=zvjcJsy51oc&t=9s

QUANTUM MECHANICS

For further investigation, check this link:
Quantum Consciousness: Bridging Quantum Mechanics and Awareness II Best Space Documentary 2024
https://www.youtube.com/watch?v=KIIJWon2ofo

Neil deGrasse Tyson & Janna Levin Answer Mind-Blowing Fan Questions
https://www.youtube.com/watch?v=kT7y1-clArQ

How Did the First Atom Form? Where did it come from? | Big Bang Nucleosynthesis
https://www.youtube.com/watch?v=4Ra9oaLouGU

The Hidden Law of the Universe: How the Least Action Principle Shapes Every Natural Phenomenon
https://www.youtube.com/watch?v=AkjHOp1myBM

THE ENTIRE HISTORY OF QUANTUM PHYSICS Explained in One Video
https://www.youtube.com/watch?v=4LychvQ-BAU

A Brief History of Quantum Mechanics - with Sean Carroll
https://www.youtube.com/watch?v=5hVmeOCJjOU

Scientists Finally X-Rayed a Single Atom. This Changes Everything
https://www.youtube.com/watch?v=uj-mjq1xshg

The Shape of A light Particle | A Quantum Breakthrough
https://www.youtube.com/watch?v=H9cFLbP251E

The Emptiness of the Universe. Immersing Deep Space
https://www.youtube.com/watch?v=ycM4NJYEniM

EXTRA DIMENSIONS

"What If You Could Access the TENTH Dimension?" | 10D Explained
https://www.youtube.com/watch?v=azpUG2GUzFI

Experiencing Higher Dimensions
https://www.youtube.com/watch?v=XAcCraWPIdQ

What Does 4D World Actually Look Like Let's Explore It with Some Examples
https://www.youtube.com/watch?v=O5ZCmfLjcrI

ALIENS

The Hidden Dangers of Discovering a Type-7 Civilization: Should We Be Worried?
https://www.youtube.com/watch?v=PW5jktHLtqQ&t=78s

Is Anybody Out There?
https://www.youtube.com/watch?v=DrSbbq4GKq0&t=600s

Brian Cox Debates If Aliens Have Visited Earth?
https://www.youtube.com/watch?v=GVkF49dbd_w

Brian Cox - Alien Life & The Dark Forest Hypothesis
https://www.youtube.com/watch?v=o6IQynhsQ1M

Alien civilizations from level 1 to level 7. We are only at level 0.72
https://www.youtube.com/watch?v=Yb4H098aMRI

Why This Advanced Civilization Has Scientists Freaked Out
https://www.youtube.com/watch?v=SqXwRNk3a8U

Alien Interview Part 2
https://www.youtube.com/watch?v=7TE6frpygVY
shows how stupid a military interrogation can be...

6 Types of Alien Life Forms Thriving in the Universe Have Terrified Scientists.
https://www.youtube.com/watch?v=Ve6eLQAJb5A

Ancient Aliens: Did The Annunaki Come from the Stars?!
https://www.youtube.com/watch?v=DrQCrLcfnZk

Brian Cox Found New Solution to The Fermi Paradox And It Isn't Good
https://www.youtube.com/watch?v=VSE_mnW-MWU

Conclusion

I've got health issues which I won't go into. But since I have them, I have to face a new reality about my writing. If I start something, I might not finish it. So, in a way, the last book I write or partly write, I must concede, must not matter in the large scheme of things.

We are a little size in the universe; and yet, we are also what the stars are made of. At the end of my life, I often think of returning to that big cycle from which I was formed...and my little life, well, will it matter?

I do not know what the answers are. I'm glad Leo Tolstoy found meaning for himself. For myself, maybe it's the act of looking. Not necessarily finding an answer or finding success but the act of trying. That might be the whole meaning of life. I wonder what the universe, if it has a purpose, is trying to do or to find? But that's a wonder and not something I have to worry about.

The exercise of writing this book has been instructive. I've learned that the universe is like a giant brain with synapses and electrons all in motion. There are sizes within sizes and it's fascinating that the outer reaches of space may be never ending. By the same token, in quantum physics, who knows how small, small really is, and whether we are only one of 11 or 29 dimensions. There is so much to learn in the Book of Nature.

Roger Penrose came up with the idea of a cyclical universe that repeatedly creates itself in a whole series going forward and going backward in time. And yet, although it's a wonderful idea, which apparently takes away the necessity for God, I still cling to the God of the

Needs, for I know how small and weak I am in all of existence.

I am grateful to the James Webb Telescope for showing me things I thought I'd never see in my lifetime. I am grateful to all the astrophysicists who are engaged in probing current secrets of the universe, and I am grateful for living in this wonderful time of discovery. I am grateful for life.

PS. A big thank you for Google and for YouTube documentaries which helped me write this book to make big ideas more accessible to the public.

Extra Thoughts

I can't remember the last time I had a dream...until this week. There are some people who insist they have no regrets in life. Well, I take that with a grain of salt. I dreamt that an old girlfriend came into the office for a passport renewal. She wore an old school jacket, light blue with the school emblem on it. I was standing in line, face averted. I spotted her; she did not spot me. When she turned to look at some brochures on the wall, I snuck out and scurried away like a scared rabbit. I suppose I was not the one to leave a relationship without hard feelings. Maybe she was a better person than I was. Strange how that goes. Who knows what my life might have been had I taken another road? Yet, I took the road I did and met my wife years later and for that, I am grateful.

The Road Taken
Something to find leads me away
To woods and lonely trails unseen
To question the road, I've taken
And the different future
That might have been

Some trail leads me
To waterfalls that crash.
My eyes see sparkles
In a single splash

John Hartig   The Universe Explained

Tiny drops, like people,
That tumble and toss
That teem and toil
And run…oh, where?
Like love, like life so fleeting
Like things, not always fair

I imagine a droplet is the sun
A splash, the galaxy
Our existence, a miracle
Our universe, you and me

Seeing is more than eyes alone
And more than merely looking
We have a mind, a heart and soul
To see the wonder in the whole

John Hartig
Monday, June 11, 2001

*****

    The recent photos that the James Webb Telescope took of distant stars and exoplanets got me thinking. I guess it involves time, space, life and death. We are lucky to be alive at this point in time when we are discovering so much, and in the face of that, when we are still playing dice with our existence on the only planet we have, our Mother Earth.

    There must be something in us which would be a shame to lose if we blew this planet apart.

    I like to think of the drive to reach the stars, to find things out, to know stuff. Maybe that's why I have the urge to write poetry and why I want people of all nations to be nicer to each other.

The Sea is Never Full
All rivers empty into the sea
Yet the sea is never full.
Is human wisdom like that?
That we float along
In the river of life,
learning skills which make us
feel grandiose and invincible
within the ocean of time!
That we hold the universe
in the palm of our hand,
that our world is more
than a mere grain of sand!
And that the human race
will never surrender
to that good night
that we, through pride defy
the dying of the light,
as if we were gods.
   Sept. 2023, John Hartig

John Hartig
Niagara Scenery + Web Design + Novels

As a photographer, I see colour, shades and patterns that the ordinary person might not see. People have busy lives and often don't take a second look.

Publication Contributions:
by John Hartig

1. Poem by John Hartig p. 43, "I Walked to Kenny's Grave Today", <u>Solitude: A Collection of New Canadian Poetry</u>, publ. 2009, Polar Expressions Publishing, Maple Ridge, BC.

2. Poem by John Hartig, "Songs of Innocence and Experience", <u>The Journey: A Collection of New Canadian Poetry</u>, publ. 20010, Polar Expressions Publishing, Maple Ridge, BC.

3. Short story by John Hartig, "Coffee Break", <u>Formation: New Canadian Short Stories</u>, publ. 2010, Polar Expressions Publishing, Maple Ridge, BC.

4. Short story by John Hartig, "Courage Getting Old", <u>From Across The River</u>, publ. 2011, Poetry Institute of Canada, Victoria B.C.

5. Center Page Photo Spread: <u>Our Canada – A Country for All Seasons</u>, "Spring Blossoms in Niagara-on-the-Lake," publ. 2012

6. Photo Design Ad Published: ARABELLA, Magazine Publication of Canadian Art, Architecture and Design, "Spring Awakenings 2012 Edition", Full page photo ad for *Granny's Boot Antiques* in Vineland, "Unique Folk-Art, Vibrant and Alive!" John Hartig Photos.

Fiction

The New Crusades, John Hartig, second ed. 2021, first publ. 2015, by Tellwell, under my penname, Waldemar Guenter, avail. through Amazon and Ingram
The New Crusades: The Sequel, John Hartig, second ed. 2021, first publ. by Friesen Press, 2016, under my pennames of Waldemar Guenter and Alexander Kucharski, avail. through Amazon and Ingram
Duplicity, publ. Amazon, 2018, John Hartig. avail. through Amazon and Ingram
Who Killed Jean-Marie Leclair? A Baroque Murder Mystery, publ. Amazon, 2019, John Hartig. avail. through Amazon and Ingram
Love and Faith Trilogy, Books I, II, III, publ. Amazon, 2019, John Hartig. avail. through Amazon and Ingram
The Polish Cowboy, publ. Amazon, 2019, John Hartig. avail. through Amazon and Ingram
The Tipperary Kid, publ. Amazon, 2019, John Hartig. avail. through Amazon and Ingram
John's Shorts: Little Stories with Big Ideas, publ. Amazon, 2022, John Hartig. avail. through Amazon and Ingram
John's Hidden Gems: Short Story Collection, publ. Amazon, 2022, John Hartig. avail. through Amazon and Ingram
Things Have Gotta Get Better Than This, publ. Amazon, 2022, John Hartig. avail. through Amazon and Ingram
The Chosen: A Violin Story, publ. Amazon, 2022, John Hartig. avail. through Amazon and Ingram
Jonah's Journey, publ. Amazon, 2022, John Hartig. avail. through Amazon and Ingram.
Johann Joachim Quantz, Gift of the Flute, publ. 2022, John Hartig, avail. through Amazon and Ingram.
The Sasquatch, Book 1, publ. 2023, John Hartig, avail. through Amazon and Barnes and Noble.

The Sasquatch: The Sequel, publ. 2023, John Hartig, avail. through Amazon and Barnes and Noble.

The Sasquatch Family: Volume 1 Human Greed and Volume 2, Redemption. Available through Amazon. 2024 John Hartig, copyright.

Time and Space: A Flight of Fancy. Publ. 2024, John Hartig, avail. through Amazon.

John's Science-Fiction: Anthology of Short Stories by John Hartig, publ. 2024. Avail. through Amazon.

Non-Fiction

Time in a Bottle Trilogy, Books I, II, III, publ. Amazon, 2019, John Hartig. avail. through Amazon and Ingram

You Love Our Milk and Honey, Book I, II, publ. Amazon, 2020, John Hartig. avail. through Amazon and Ingram

The Second Wave: Living Through Trump and Covid, publ. Amazon, 2021, John Hartig. avail. through Amazon and Ingram.

77 Looking Back: My Sort of Diary 2022-2023, publ. Amazon, 2023, John Hartig,

Other

Can You Imagine? A children's picture book with poetry, publ. Amazon, 2019, John Hartig. avail. through Amazon and Ingram

Poetry Like Raindrops, publ. Amazon, 2019, John Hartig. avail. through Amazon and Ingram

Battle of the Violins, publ. Amazon, 2019, John Hartig. avail. through Amazon and Ingram

John's Photobook Series, Ball's Falls to Niagara Falls, publ. Amazon, 2021, John Hartig Photos. avail. through Amazon and Ingram

Louis Riel and Me, publ. Amazon, 2021, John Hartig, a historical fiction. avail. through Amazon and Ingram

Give Us Hopes and Dreams, publ. Amazon, 2021, John Hartig. avail. through Amazon and Ingram

Where Do Good Atheists Go? publ. Amazon, 2021, John Hartig. avail. through Amazon and Ingram

We Are Not Alone: Civilizations in Outer Space publ. Amazon, 2022, John Hartig. avail. through Amazon and Ingram.

The Cosmos: Origins and Aliens, publ. Amazon, 2022, John Hartig. avail. through Amazon and Ingram.

Johann Joachim Quantz, publ. Amazon, 2022, John Hartig, avail. through Amazon.

Two Baroque Prodigies, "Quantz flute tutor for Frederick the Great, Leclair, violinist murdered in 1764", publ. Amazon, 2022, John Hartig, also avail. in English, French and German.

The Murderous Sea: Gateway to Freedom, publ. Amazon, 2024, John Hartig, avail. in English, French and German.

The Universe Explained, well sort of... 2024, publ. Amazon, John Hartig

John's Photobook Series

Besides writing, my other passion is photography. I love taking scenery pictures, especially flowers and sunsets. I want to share my pretty pictures with the public through John's Photobook Series.

Photobooks are 8.5x8.5"

The Niagara Peninsula

The Bruce Trail

Ball's Falls

Port Dalhousie

The War of 1812

Granny's Boot Antiques

Sights in the Niagara Peninsula

Niagara-on-the-Lake

Morningstar Mill

Niagara Falls

5 Waterfalls in Niagara

Fair Havens

Two of my Favourite Seasons

70th Anniversary

Some of My Best

From The Bottom Up

My Choicest Picks 1

My Choicest Picks 2

*****

- John Hartig Novels through Amazon, google title and John Hartig
- John's Photobook Series ordered directly from Amazon.
- Prints, any size enlargements, e-mail John directly to place an order. Pickup at the house, otherwise + shipping cost

John Hartig is a Niagara scenery photographer/ web designer and novelist.

John lives in Vineland, Ontario. His Photobooks and photo prints are available for home or office.

CONTACT
Ph: 905-562-7821
johnehartig@gmail.com
Websites
niagarafinearts.org
johnhartig.ca
niagarascenery.com
Or just Google
John Hartig
Photography

www.ingramcontent.com/pod-product-compliance
Lightning Source LLC
Chambersburg PA
CBHW050307230526
45471CB00005B/2062